EXPERIMENTAL INQUIRIES

AUSTRALASIAN STUDIES IN HISTORY AND PHILOSOPHY OF SCIENCE

VOLUME 8

The titles published in this series are listed at the end of this volume.

EXPERIMENTAL INQUIRIES

Historical, Philosophical and Social Studies of Experimentation in Science

Edited by

H. E. LE GRAND

Department of History and Philosopy of Science,
University of Melbourne

KLUWER ACADEMIC PUBLISHERS

DORDRECHT / BOSTON / LONDON

Library of Congress Cataloging in Publication Data

Experimental inquiries : historical, philosophical and social studies
of experimentation in science / edited by Home Le Grand.
p. cm. -- (Australasian studies in history and philosophy of
science ; v. 18)
Includes bibliographical references and index.
ISBN (invalid) 07923070909 (alk. paper)
1. Science--Experiments--History. 2. Science--Experiments-
-Philosophy. 3. Science--Experiments--Social aspects. I. LeGrand,
H.E. (Homer Eugene), 1944- . II. Series.
Q182.3.E96 1990
507.2--dc20 90-4571

ISBN 0-7923-0790-9

Published by Kluwer Academic Publishers,
P.O. Box 17, 3300 AA Dordrecht, The Netherlands.

Kluwer Academic Publishers incorporates
the publishing programmes of
D. Reidel, Martinus Nijhoff, Dr W. Junk and MTP Press.

Sold and distributed in the U.S.A. and Canada
by Kluwer Academic Publishers,
101 Philip Drive, Norwell, MA 02061, U.S.A.

In all other countries, sold and distributed
by Kluwer Academic Publishers Group,
P.O. Box 322, 3300 AH Dordrecht, The Netherlands.

Printed on acid-free paper

Printed in the Netherlands

TABLE OF CONTENTS

FOREWORD

The institutionalization of History and Philosophy of Science as a distinct field of scholarly endeavour began comparatively early -- though not always under that name -- in the Australasian region. An initial lecturing appointment was made at the University of Melbourne immediately after the Second World War, in 1946, and other appointments followed as the subject underwent an expansion during the 1950s and 1960s similar to that which took place in other parts of the world. Today there are major Departments at the University of Melbourne, the University of New South Wales and the University of Wollongong, and smaller groups active in many other parts of Australia and in New Zealand.

'Australasian Studies in History and Philosophy of Science' aims to provide a distinctive publication outlet for Australian and New Zealand scholars working in the general area of history, philosophy and social studies of science. Each volume comprises a group of essays on a connected theme, edited by an Australian or a New Zealander with special expertise in that particular area. Papers address general issues, however, rather than local ones; parochial topics are avoided. Furthermore, though in each volume a majority of the contributors is from Australia or New Zealand, contributions from elsewhere are by no means ruled out. Quite the reverse, in fact -- they are actively encouraged wherever appropriate to the balance of the volume in question.

R. W. Home
General Editor
Australasian Studies in History and Philosophy of Science

H.E. LE GRAND

INTRODUCTION: SETTING UP THE APPARATUS

Scientists actively engage in designing and assembling experimental set-ups, conduct experiments, record experimental results and produce publications presenting those results. Controversies over theories often appear to centre upon the adequacy and validity of experiment-based data. Laboratory work remains a distinctive feature of science pedagogy and textbooks are often packed with examples, illustrations and arguments drawn from experimental practice. It would seem that in any account of science, especially a naturalistic one, experimentation should loom very large. Yet, for much of the past half-century the writings of many historians, philosophers and sociologists of science have been marked by what has been aptly termed by Allan Franklin (1986) 'the neglect of experiment'.

Historians of science, especially since the appearance in 1962 of Kuhn's *Structure of Scientific Revolutions*, have tended to concentrate on what were seen to be larger issues of theory change. For Kuhn (1962: 27-34), in 'normal science' the roles of experimentation were confined largely to the improvement of theory-fact matching or to the more precise determination of paradigm-significant experimental facts. Only occasionally, in periods of 'crisis', did experimentation appear to play a more constitutive role but 'since no experiment can be conceived without some sort of theory, the scientist in crisis will constantly try to generate speculative theories...'(87). Even those historians critical of Kuhn's general schema did not, as a rule, disagree with his emphasis on theory. As Lenoir (1988: 3) has recently put it, '...the history of science is almost always written as the history of theory'.

Philosophers of science, with the notable exceptions of Hacking (1983) and a few others, have taken 'science' to be almost co-extensive with 'theory'. In Hacking's view (1983: 149-51), experimentation sometimes 'has a life of its own' and the relations between theory and experimentation can take diverse forms (159): some theories 'languish' because of a lack of connection through experiment with the 'real' world; some experimental phenomena 'sit idle' for lack of appropriate theory. But, for most contemporary philosophers, devising, testing, modifying, comparing and choosing theories have been the staple of innumerable articles and monographs. Experiment as part of scientific practice has generally been regarded as unproblematic and as subsidiary to more crucial issues. Experiment, for example, is regarded as an onerous but necessary step to gather data, in an unproblematic way, which furnishes the raw materials for the important work of the theorist. Or, experiment may be used, again in an unproblematic way, to 'test' a theory or to referee among rival theories. Popper, perhaps, has a somewhat greater role for experiment but in his view of science, the primary reason for constructing experiments is as a straightforward attempt to refute a theoretical 'conjecture'. Even for those concerned with such issues as the Duhem-Quine thesis, *theory* has been the centre of attention and *experiment* seems to have been regarded as philosophically uninteresting.

Sociologists of scientific knowledge of various persuasions, most notably Harry Collins (1985), have sought with a fair measure of success to problematize experiment. In the most radical form, the argument goes well beyond the Duhem-Quine thesis and the theory-ladenness of observations to claim that a presumed external world places no constraints on scientific practice. The authentication, acceptance and interpretation of experimental results are the outcome of social negotiation. Experiment, except as a potential site for studying scientific controversy, is so problematized as to be sociologically trivial and uninteresting.

There have, however, recently been reassertions of the importance of experiment from philosophers, historians and sociologists (Franklin, 1986; Galison, 1987; Latour, 1987; *Science in Context*, 1988, 2; Giere, 1988; Batens & van Bendegem, 1988; Gooding, Pinch & Schaffer, 1989). The positions taken acknowledge the contingency and negotiation involved in 'doing', as well as 'interpreting', experiments. These range from the view that experiments are, at least on occasion, of decisive importance in opening and closing scientists' controversies about 'Nature' (*e.g.*, Giere, 1988: 165-167, 198-199) to the more minimalist position that experimental results or 'facts', the outcome of 'trials', furnish powerful allies for competitors on the agonistic field of science (Latour, 1987).

The studies in this volume complement and extend the recent, growing body of analyses of the multifarious roles of experiment in science. All share a naturalistic approach in that each is attentive to the interpretative and argumentative practices of scientists. Yet, each has a different focus, both in terms of experimentation *per se* and in terms of scientific subject matter. Some of the contributors are concerned primarily with recovering the cognitive and social contexts in which specific experiments were conducted and interpreted, others with the use of experiment as an argumentative resource, still others with the variable linkages between experiment and theory. The experiments surveyed extend from those of the era of the 'Scientific Revolution' of the sixteenth and seventeenth centuries down to the 1980s and includes examples drawn from not only physics, chemistry and biology but also anthropology, geology and medicine.

John Schuster and Graeme Watchirs address the 'eighteenth century problem' in terms of the relationship between the older enterprise of natural philosophy and the various experimental sciences constituted near the turn of the nineteenth century. They reject the view that these new domains came into being through some sort of epistemological rupture (Bachelard) or acquisition of first paradigm (Kuhn). Instead, they argue that the study of early experimental science should proceed from an understanding of the social and cognitive nature of natural philosophy and its shift into, and the internal dynamics of, *experimental natural philosophy*. The transformation of natural philosophy into experimental natural philosophy, *ca.* 1650, involved a reconfiguration of the declared meaning and place of experience and experiment. The dynamics of experimental natural philosophy involved attempts to embody natural philosophical discourse within experimental hardwares.

Schuster and Watchirs illustrate their thesis with the case of electrostatics, focusing upon Franklin and the Leyden jar. In their analysis of experimentation in electrostatics they develop the notion of a 'hardware-discourse couple', of which the Leyden jar is one example. In accord with Bachelard's conception of the 'phenomeno-technique', they contend that experimental hardwares embody conceptual structures but, unlike Bachelard, the meanings thus embodied are not 'frozen' but are fluid and revisable. Hardwares and the discourses in which they are embedded are contingent and subject to renegotiation on either side of the 'couple'. As they put it (23-24), "Instrumental and experimental hardwares are accountably 'adequate' materializations of theories and concepts argued/enforced as 'adequate' at some point in the past and 'until further notice'." In their view, the discursive resources which Franklin was striving to embody in such hardware as the Leyden jar were drawn from natural philosophy. The theories and concepts he was endeavouring to thereby 'black box' were qualitative and therefore did not include inscription devices. That development was later associated with the quantification of electrostatics and the transformation of electrostatics from a domain of experimental natural philosophy to a domain of experimental science. From this perspective, Franklin's understanding of the Leyden jar represents an exemplary case of the continuities in and dynamics of experimental natural philosophy, not a rupture with previous practices and understandings.

Bruno Latour's programmatic essay sets experiment in the context of actor-network analyses of science and addresses both 'realist' and 'social constructionist' construals of experiment. An experiment is a test of the strength, a trial, of associations which are at once social and cognitive, political and technical. Actors, both human and non-human, are defined in terms of the trials that they have undergone. Experiment is directed to 'Nature' but also to one's colleagues, competitors and the scientific literature. In an experiment there is a 'trial', an 'inscription' (*i.e.*, a written record, the element which distinguishes 'being experienced' from 'performing an experiment') and a point to be made or unmade in the literature. It is a powerful resource in controversies: as Latour (59) phrases it, experiment introduces allies 'that are non-political, that are reliable, and that put discussions to an end.' Experiment is both artificial and real. It is artificial in that experimental 'facts' are 'fabricated' in laboratories yet they are 'real' in that they have gained definition by *passing* and *continuing to pass* trials. As a result, 'there is a history of science, not only of scientists, and there is a history of things, not only of science' (66).

Latour exemplifies his view that experiment is a collective and continuous historical event with a discussion of Baudin's mapping of Tasmania. This is, as Latour notes, a paradoxical case of 'experiment' but it is also a challenging 'trial' for his programme: how could one possibly think of Tasmania as an 'actor'; how can one argue that Tasmania becomes defined through a series of trials? Latour's answers reflect scepticism of naive realist and social constructionist approaches. The shape of 'Tasmania', he concludes is a collective construction yet it is 'out there' as visitors from Baudin to present-day tourists affirm (thereby maintaining its stability). Latour also addresses the issue of cultural relativism through his consideration of a public 'experiment' conducted by Moudang diviners in Chad. He points to similarities between their 'ritual' and the western-style 'experiments' conducted by, *e.g.*, Archimedes, Baudin and Pasteur but he also identifies a crucial difference. In the former, relations of forces *are traced out* by *quasi-objects*; in the latter, relations of forces *are reversed* by *objects*. In the former, there is a monism -- one single society; in the latter, a dualism because scientists can enforce a distinction between *science*, *i.e.*, the non-human (phenomena, facts, instruments, yeast, coastlines, levers) and *society*, *i.e.*, the human (politics, psychology). It is in this sense that experimentation has a special character: it can be used as a lever to alter or reverse the balance of social and cognitive forces taken together.

Miranda Hughes examines F.A. Peron's experiments with the dynamometer. Peron, a self-styled 'anthropologiste', sailed on Baudin's expedition to the southern Pacific at the turn of the nineteenth century. His measurements of what he took to be the relative strengths of Tasmanian aborigines, other Pacific peoples, English settlers and French sailors gave him a lever with which he overturned Enlightenment assumptions about the physical inferiority of 'Civilized Man' to 'Savage Man'. In Schuster and Watchirs' terms, his investigations constitute an instance of 'experimental science' rather than 'experimental natural philosophy'. The discourse upon which he drew was a traditional natural philosophical one but that qualitative discourse was transformed by the quantitative hardware, the dynamometer, with which he coupled it. The reshaped discourse was directed toward an (objective) science of man and an (objective) science of work.

Hughes regales the reader with an account of Peron's sometimes amusing errors, the difficulties he encountered in trying to secure co-operation from his 'experimental subjects', the lack of experimental protocols, the paucity of his data, the theory-ladenness of his results and conclusions. All this makes even more dramatic the power of experiment -- its ability through Peron to, as Latour puts it, 'reverse the balance of forces'. The laboratory had sailed to the southern seas and it spoke through Peron alone. His measurements on eighteen 'sauvages' read off the (wrong!) dial of an objective instrument and inscribed in his tables of results in the end counted for more than the opinions, the theories, the collected casual observations and sailor' stories of all those who had remained behind in France. The insertion of the dynamometer between the 'anthropologiste' and his subjects rendered in the judgment of Peron's contemporaries science out of what had been speculation and inverted the previously accepted inverse relationship between strength and civilization.

Jan Golinski directs our attention to the experimenter's audience. He argues, in common with Latour and many others, that experiment can serve as a 'lever' to move the opinion of the experimenter's audience. The audience, whether composed of specialists or members of a more general public, whether actual or 'virtual' (Shapin, 1984) witnesses of the experiment, can thereby be enlisted as a resource in controversies over knowledge claims. Golinski chooses two episodes in the career of Humphry Davy to illustrate the importance and complexities of the interaction between experimenter and audience.

The first episode centres upon Davy's efforts to establish the character and reliability of the voltaic pile as an instrument of chemical analysis. As Golinski points out, this involved both removing the pile from natural philosophical discourse about subtle fluids to a technical discourse about tools of chemical analysis and converting the pile from an 'experiment' into an 'instrument'. Davy pursued this goal through an array of technical or, as Franklin (in Gooding *et al.*, 1989: 438) terms them, 'epistemological' strategies: the elimination of alternative explanations, systematically removing possible sources of error, varying the experimental set-up and so on. At the same time, Davy adopted a social strategy of enlisting both a public audience through his demonstration lectures at the Royal Institution and a more elite audience through his Bakerian Lectures to the Royal Society. In this first episode Davy was highly successful in his goal of transforming the pile into a reliable transmitter of 'Nature'; indeed, so successful was he that, perhaps to his chagrin, some members of his audience attributed his discoveries to the instrument rather than to the experimenter. In the second episode, the controversy over whether or not oxymuriatic acid [chlorine] was a simple body [element], Davy was less successful. He encountered resistance in his 'social strategy' of attempting to enlist and 'discipline', to use Latour's term, an audience of specialists. Specialists had sufficient resources on occasion to dispute Davy's exegisis of experiments and to provide alternative interpretations. This is, of course, connected with the limitations of Davy's 'technical strategies' in this case: he had no dramatically new experimental evidence to present, he produced no new substances to exhibit, and his contention that chlorine was an element went against generally accepted chemical theory. Golinski's study brings out clearly not only the intimate relationships among experimenter, experiment and audience but also underlines the difficulties which can be encountered in mobilizing experiments and audiences, separately or together, to open or close disputes.

Jan Sapp explores the ways in which scientists construct and reconstruct the meaning of and motivation for previous experiments as a resource for their own work. He takes as exemplary the extraordinary variety of conflicting accounts by scientists and science analysts of Mendel's legendary experiments. The 'long neglect' and 'founding father' myths which cluster around Mendel reflect not so much a naive whiggism as the use of history to open, sustain or close later controversies. In particular, he examines how Mendel's [imputed] motives for, interpretation of, and significance ascribed to, his famous pea experiments were constructed and reconstructed in the context of twentieth-century debates and research programmes in biology. These conflicting accounts are not a result of 'misreading' Mendel. Rather, Sapp suggests, some of the reasons why Mendel has been so widely cited, has been so 'influential', and has attracted so much attention from scientists (as well as science analysts) are the very ambiguities in and scantiness of his published papers.

Mendel and his experiments were construed by Bateson, for example, so as to ally them with experimental genetics and non-Darwinian evolution in Bateson's struggle with statisticians. Fisher, a statistician and an architect of the 'modern [Darwinian] synthesis', contended to the contrary that Mendel and his experiments were part of the Darwinian tradition. Similarly, the 'historical' controversy over whether or not Mendel *really* obtained his claimed experimental results can be construed in terms of a later methodological debate. Experimentalists might claim that Mendel's 'laws' were just generalizations from his data; non-experimentalists, that Mendel's experiments, if performed at all, were intended as demonstrations of 'laws' arrived at on other grounds. Sapp, in the course of his study, repeatedly shows that scientists themselves were keenly aware -- well before historians, philosophers and social analysts of science -- of such issues as the theory-ladennes of observations, the underdetermination of theories by data, observer bias and the rhetorical character of scientific papers. He also shows that not only experiments, but also the history surrounding them, are malleable, revisable and negotiable.

David Turnbull and Terry Stokes survey contemporary biomedical experimentation as practiced in the Walter and Eliza Hall Institute. Viewing research as a labour process, they argue that a key element is the establishment of a 'manipulable system' for the generation of data. A 'manipulable system' is made up of 'instruments, skills, techniques and raw materials, together with theories, all embedded in social relations' (169; *cf.* Latour's 'black boxes' and Schuster and Watchirs' 'hardware-discourse couples'). They, like other science analysts (*e.g.*, Gooding *et al.*, 1989: 2-5, 183-223; Hacking, 1983: 220, 230), point out that much of the work of experimenters is getting such assemblages to work, not only in the experimenters' judgment but also in that of their colleagues. Turnbull and Stokes contend that the power of science flows from the material conditions and practices embodied in such systems (*cf.* Gooding *et al.*, 1989: 9; Hacking, 1983: 262; Galison, 1987; Giere, 1988: 5, 113, 139).

Turnbull and Stokes take two examples of systems in operation at the Hall Institute. As an example of the difficulties and contingencies involved in establishing a manipulable system, they describe the setting up of the 'Malaria System' constructed to isolate antigens from malaria parasites. As an example of the manipulation of a system to produce a published 'experiment', they recount the process by which the 'Thymocyte System' was built and then used to generate an experimental report.

Stephen Gaukroger takes an explicitly philosophical approach to the relationship between experiment and theory. In particular, he rejects the view that the role of experiment is limited to determining the 'truth' or 'falsity' of theories. His premise is that theories about the meaning of scientific statements underlie and are prior to accounts of the role of experiment. He sketches a taxonomy of theories of meaning and, drawing upon Michael Dummett, argues in favour of a 'moleculist' theory, at least for the discovery and articulation of new theories. Moleculism involves a 'double mode of presentation' with a linguistic *and* an extralinguistic dimension: one the one hand, grasping the sense of a scientific statement; on the other, identifying to what it is that the statement refers. Experiment provides a route to the latter; indeed, it can provide an empirical route, additional and alternative to the conceptual route, to identifying the referent of a statement. Gaukroger illustrates the advantages of a moleculist account in a reconsideration of Galileo's formulation of his law of falling bodies.

Andy Pickering opens up for analysis the question of how scientists, experimentalists and non-experimentalists alike, establish goals for their work: how and why do they decide to extend scientific culture in one direction rather than another? He develops this theme through an analysis of Luis Alvarez's formulation of his goal of a large hydrogen bubble chamber and his implementation of plans to reach that goal. He suggests that Alvarez's goal was a projection into the future of imaginatively constructed intersections among heterogeneous elements of scientific culture: *e.g.*, a small liquid-filled chamber developed by Donald Glaser, theories of elementary-particle interactions, the notion of the hydrogen atom as a target, Alvarez's experience in 'big science' physics, the financial and engineering resources Alvarez had at his disposal. These intersections helped close off alternative directions and helped stabilize Alavarez's goal but, as Pickering shows, Alvarez's goal and his plans for achieving it were not fully precise and unrevisable until the big bubble chamber 'worked'. Between the formulation of the goal and its realization were various obstacles, 'situated resistances', opportunistic decisions and serendipitous occurences which could, and sometimes did, lead to revisions of his goal.

Pickering sets his study in the broader context of a pragmatist model of scientific practice. He has, however, sought to extend the process of goal formation beyond the 'routinized goals of daily life' (231) to encompass scientists' goals. He invokes for this purpose the importance of modelling: as he puts it, 'goals as future states are built out of existing cultural elements, analogically transformed' (217).

Homer Le Grand sketches arguments for the importance of visual presentations in science. He contends that pictures, diagrams and other forms of visual representation of both data and theories can be powerful tools for the dissemination and the concentration of knowledge and knowledge claims and for the mobilization of resources in controversies. He 'illustrates' his arguments with 'symbolic maps' and 'symbolized theories' constructed and deployed in the context of the recent controversy over theories of continental drift. More provocatively, he argues that, at least in the earth sciences, illustrations have many analogies with and sometimes fill the traditional roles of experiment. He closes by pointing to recent work in the field of cognitive science which suggests why 'pictures' may have such evident power and importance. This could provide a new perspective on why 'seeing is believing', not only in the case of 'pictures' but also, more pointedly, in the case of experimenters and their audiences.

It will be plain to the reader that more questions are posed in this volume than are answered; that the topic of experimentation invites much further research. Nonetheless, it is also plain that some questions have been more clearly defined and some additional ones identified; that some useful analytic tools have been forged and some more familiar ones re-sharpened; and that there is a measure of consensus on the outlines of some answers.

REFERENCES

Batens, D. & van Bendegem, J.P., eds. (1988). *Theory and experiment*. Synthese Library vol. 195, Dordrecht: Kluwer

Collins, H.M. (1985). *Changing order*. Beverley Hills: Sage.

Franklin, A. (1986). *The neglect of experiment*. Cambridge: Cambridge University Press.

Galison, P. (1987). *How experiments end*. Chicago: University of Chicago Press.

Giere, Ronald N. (1988). *Explaining science: A cognitive approach*. Chicago: University of Chicago Press.

Gooding, D., Pinch, T. & Schaffer, S., eds. (1989). *The uses of experiment: Studies in the natural sciences*. Cambridge: Cambridge University Press.

Hacking, I. (1983). *Representing and intervening*. Cambridge: Cambridge University Press.
Latour, B. (1987). *Science in action*. Cambridge: Harvard University Press.
Lenoir, T. (1988). Practice, reason, context: The dialogue between theory and experiment. *Science in Context*, **2**, 3-22.
Shapin, S. (1984). Pump and circumstance: Robert Boyle's literary technology. *Social studies of science*, **14**, 481-520.

University of Melbourne

JOHN A. SCHUSTER & GRAEME WATCHIRS

NATURAL PHILOSOPHY, EXPERIMENT AND DISCOURSE: BEYOND THE KUHN/BACHELARD PROBLEMATIC

1. INTRODUCTION

The eighteenth century witnessed the constitution of mature domains of experimental research in chemistry, physical optics, electricity and magnetism, and heat theory. This poses one version of the problem of the eighteenth century in the history of science: What is the shape of the period between the end of the Scientific Revolution and the 'Second Scientific Revolution', that era in the late eighteenth and early nineteenth centuries that saw the accelerated development of experimental fields, the mathematization of some, and the development of their institutional and increasingly professional structures? Attention has been lavished upon other related construals of the eighteenth-century problem, notably the meaning and scope of Newtonianism, the relations between science and the Enlightenment, and the exigencies of eighteenth-century matter theory including, recently, its social relations. But, the maturation of new experimental or mathematico- experimental fields remains fundamental for our understanding of the shape of the history of European science, theoretical perspectives on discipline formation, and the nature of experiment.

Gaston Bachelard and Thomas Kuhn recognized the historiographical import of the eighteenth-century maturation of fields of experimental science and their respective models of theory dynamics were conditioned by this problem. The influence exerted by their works alone justifies a reconsideration of their views. This conclusion is only reinforced when one further considers that Kuhn's problematic lingers on, shaping and constraining even the best Anglo-American historiography of eighteenth-century science and that Bachelard's account of what constitutes experimental sciences bears some relations of analogy and paternity to current sociological perspectives on experiment. However, today the problem of eighteenth-century experimental fields has even greater salience due to the confluence of two strands of research in the history of science and the sociology of

H. E. Le Grand (ed.), Experimental Inquiries, 1–48.

scientific knowledge. On the one hand, there has been renewed, if somewhat inchoate, attention to the category of 'natural philosophy' as a key to understanding the eighteenth century (Schaffer, 1980; 1983; 1986; Shapin, 1980; 1981; Shapin & Schaffer, 1985; Wilde, 1982; Cantor, 1982). Neither Kuhn nor Bachelard had a coherent and positive notion of eighteenth-century natural philosophy. On the other hand, sociologists of scientific knowledge are increasingly attentive to the socio-politics of experiment and scientific instruments, and the eighteenth century is a key site for applying and refining social constructivist perspectives hitherto forged mainly, though not entirely, by considering contemporary cases.

This paper triangulates amongst three elements: the Kuhn/Bachelard problematic of eighteenth-century experimental fields, the renewed awareness of natural philosophy as an historical category, and the emerging social constructivist literature on experiment. Our chief aim is to sketch a post-Kuhn/Bachelard framework for analyzing and describing the nature and dynamics of the nascent eighteenth-century domains. Our triangulation thus has a definite focus and point of application, for this effort turns on the issue of understanding 'natural philosophy', especially the experimental natural philosophy of the seventeenth and eighteenth centuries, as an historical enterprise and as an historiographical category. It is in this area that we hope to clarify the eighteenth-century problem and connect our solution to historiographical issues about the preceding 'Scientific Revolution' as well. Nevertheless, our other two points of reference are inextricably involved as sources of insight and as objects of critical discussion. First, by establishing a preliminary model of the character and dynamics of experimental natural philosophy, we can turn the Kuhn/Bachelard problematic on its head whilst specifying the standing of their respective contributions. Kuhn and Bachelard practise a 'two-place' historiography, seeking points of origin at which eighteenth-century experimental sciences emerged from pre- or non-scientific enterprises. In contrast we shall argue there are no such science-founding events or achievements: these are misunderstandings of moments in the social and technical dynamics of experimental natural philosophy. Still, we will reach this post-Kuhn/Bachelard conclusion partly with the help of their own analyses, thereby contributing to a determination of the current standing of their views on the history of experimental sciences and science dynamics generally. Second, since our model of experimental natural philosophy will also depend in part upon adapting perspectives in the literature on the sociology of experiment, we will in effect address the question of the pertinence of historical analyses to the sociology of scientific knowledge. The chief example used to illustrate our argument will be the development of electrostatics in the eighteenth century, and especially the role of the Leyden jar and of Franklin's work in that development: an area of eighteenth-century 'science' that had considerable importance in the deliberations of Bachelard and Kuhn.

Our argument involves 'tacking' among the three elements mentioned above and it is, of necessity, rather complex. Sections 2 and 3 examine Kuhn's and Bachelard's two-place historiographies of eighteenth-century fields, with special attention to the case of electrostatics. Section 4 explores the sense in which Kuhn, whilst lacking a positive conception of natural philosophy, was nevertheless groping toward some sense of continuity across the pre-paradigm/paradigm divide, a continuity we suggest can only have been in and of natural philosophy. Then, in Section 5, we articulate the category of natural philosophy, developing a working conception of experimental natural philosophy that requires further development in the form of an explicit model of experiment and experimental hardware as forms of embodied discourse. This model, essential for an understanding of experimental natural philosophy, is developed in Section 6 by 'crossing' our analysis of Bachelard with some perspectives in the constructivist sociology of experiment. This in turn makes possible the treatment of experiment within a refined model of experimental natural philosophy, and Section 7 develops this in the light of the aspects of electrostatics treated by Kuhn and Bachelard. Section 8 offers an overview of the dynamics of experimental natural philosophy. Finally, Section 9 draws some conclusions for the history of science and natural philosophy.

2. KUHN: THE GENESIS OF EXPERIMENTAL FIELDS

Both Kuhn ad Bachelard practiced a 'two-place' historiography; *i.e.*, both seek ruptures or sudden births, where sciences are born, placing such events in the context of the pre-existence of what is 'not-science'. Kuhn identifies the genesis of an experimental science with the emergence of a first 'paradigm', against a field of 'pre-paradigmatic' enterprises in the relevant domain of research. Bachelard identifies the emergence of an experimental field with the sudden creation of a 'science' that replaces 'pre-scientific' enterprises or discourses. Neither Kuhn nor Bachelard seriously entertains 'natural philosophy' as an historical category and so neither of them can discern the possibility that what they call sciences or first paradigms actually emerged as a result of the characteristic dynamics of experimental natural philosophy.

At the beginning of *The Structure of Scientific Revolutions*, Kuhn uses electrostatics to illustrate his theory of paradigm formation. The pre-paradigm stage of electrical research was dominated by competing schools with differing metaphysical commitments (*e.g.*, Cartesian or Newtonian), preferred problems, and starting points for surveying the burgeoning field of known electrical phenomena (*e.g.*, stressing the phenomena of attraction/repulsion rather than of conduction). For Kuhn, the founder of the first real electrical paradigm was Benjamin Franklin, initially an adherent of the pre-paradigm school concerned mainly with conduction effects. This school 'tended to speak of electricity as a "fluid" that could run through conductors, rather than as an effluvium that emanated from non-conductors'. They could deal with simple conduction effects, but not very well with the known attraction/repulsion effects. The work of Franklin and his successors, however, issued in a successful first paradigm 'that could account with something like equal facility for very nearly all these effects and therefore [provided] a subsequent generation of electricians with a common paradigm for its research' (Kuhn, 1970: 14-5).

The 'exemplar' of the Franklinists, their exemplary problem solution out of which their full paradigm was articulated, was their 'successful' explanation of the Leyden jar. In whiggish terms the jar was a primitive form of condenser. It might typically consist of a glass container filled with water and stopped with a cork . A metal pin was inserted through the cork into the water. The jar was charged by grounding the outside surface of the glass and bringing the pin into contact with a prime conductor deriving electrical fluid from the rotating globe of an electrostatic generating machine via its collecting brushes and chain. If the circuit between the pin and the outer surface were closed, the jar would discharge. The jar was devised in an attempt to store electrical fluid, but it was found to work effectively only when the usual rule for charging 'non-electrics' (derived from Dufay) was not followed; that is, the jar had to be grounded, rather than insulated in order to be charged. According to Franklin, the electrical fluid added to the inside of the jar repels an equal quantity of fluid off the jar's outside surface. In discharge to neutrality, the surcharge within the jar flows out through the pin and restores to the outer surface a quantity of fluid equal to that originally repelled off that surface.

According to Kuhn, to understand the basis of Franklin's exemplar one must look back to the origins of the pre-paradigm 'fluid' school of electricians. Only in the 1740s had it become legitimate to invoke imponderable, aether-like fluids: weightless, property-bearing, conserved and capable of acting at a distance. Kuhn (1970: 106) argues that for aether- or fluid-theorists,

> electrical phenomena increasingly displayed an order different from the one they had shown when viewed as the effects of a mechanical effluvium that could act only by contact. In particular...the phenomenon we now call charging by induction could be recognized. Previously, when seen at all, it had been attributed to the direct action of electrical 'atmospheres' or to the leakages inevitable in any electrical laboratory.

It was this possibility of conceptualizing electrostatic induction, and hence the possibility of distinguishing between, on the one hand, the flow and conduction of electrical fluid and, on the other hand, its exertion (upon itself) of repulsion at a distance, that, according to Kuhn, facilitated Franklin's paradigm-making achievement of explaining the Leyden jar (1970: 117-8). Indeed, for Kuhn it was these insights that constituted the essence of the Franklinian paradigm, as he makes quite clear (1970: 118) by conflating the establishment of that paradigm with the widespread comprehension and acceptance of his clarification of inductive effects and of conduction versus distance effects:

> ... after the assimilation of Franklin's paradigm, the electrician looking at a Leyden jar saw something different from what he had seen before. The device had become a condenser, for which neither the jar shape nor glass was required. Instead, the two conducting coatings--one of which had been no part of the original device--emerged to prominence. As both written discussions and pictorial representations gradually attest, two metal plates with a non-conductor between them had become the prototype for the class. Simultaneously, other inductive effects received new description, and still others were noted for the first time.

By packing this understanding of what we might term 'the condenser as such' into the paradigm from the point of its 'assimilation', Kuhn endows Franklin's paradigm with a destiny of unproblematically maturing through articulation of its own pre-given conceptual and technical resources (1970: 21-2, 118; 1963: 356-7).[1] We shall see in Section VII that Kuhn's account of the core of Franklin's exemplary achievement is quite misleading. For the moment we shall take Kuhn's claims at face value and examine his 'two-place' historiographical use of the concept of a 'first paradigm' in the case of electrostatics.

On the one hand, Kuhn stresses the rupture with the ineffective, not really cumulative, indeed not scientific pre-paradigm state; a rupture pinned upon the Franklinians having achieved clarity about induction. On the other hand, once the rupture has occurred, an epistemological predestination hovers around Franklin's paradigm. Assuming, as Kuhn does, that Franklin was clear about distinguishing induction from conduction, and, in general, flow effects of the electrical fluid from its distance effects, then a Kuhnian can gaze almost whiggishly at the cumulative articulation of Franklin's paradigm. One learns (1970: 18):

> What the fluid theory of electricity did for the sub-group that held it, the Franklinian paradigm later did for the entire group of electricians. It suggested which ex eriments would be worth performing and which, because directed to secondary or overly complex manifestations of electricity, would not. Only the paradigm did the job far more effectively, partly because the end of interschool debate ended the constant reiteration of fundamentals and partly because the confidence that they were on the right track encouraged scientists to undertake more precise, esoteric, and consuming sorts of work. Freed from the concern with any and all electrical phenomena, the united group of electricians could pursue selected phenomena in far more detail, designing special equipment for the task and employing it more stubbornly and systematically than electricians had ever done before.

Given that the Franklinians had their exemplar well sorted out, Kuhn can see the maturation and articulation of Franklin's paradigm as the progressive mathematization of electrostatics (1970: 21, 28-9; 1977: 47-8; 1963: 356-7). Here, Kuhn's historiography partially echoes whiggish histories of electrical science. Garden variety whiggism weaves a story of the clarification and generalization of facts and the instrumentalist rejection of the imponderable electric fluid; for Kuhn, mathematization is made possible by and played out in the Franklinian paradigm.[2] Hence Kuhn, like ordinary positivist whigs, is ultimately a seeker after 'scientificity'. Both historiographies are dramas of science versus non-science. But, there is one telling difference. Whig history can afford to see a science maturing gradually, because it attends to the accrual, generalization and perfection of 'facts'. The drama is long and slow. Kuhn's catastrophism of conceptual frameworks, inherited from the historiography of Koyre, inoculated him against naive empiricism and led him to the notion of a *first* paradigm in a field, and thence to a history of discontinuity since first paradigms constitute ruptures with the non-scientific past *and* pre-program the maturation of their field.[3] The net result, therefore, is a dramatic two-place historiography, of early electrostatics and other experimental fields, premised on locating the first outbreak of scientificity.

The inadequacies of Kuhn's two-place historiography have been highlighted by recent historical research on electrostatics, some partially indebted to Kuhn. Consider the path to mathematization programmed into the paradigm by the exemplary 'condenser as such'. The work of

Heilbron, Home and others has well shown that Franklin had no modern text-book conception of induction. Moreover, it is clear that Franklin, committed as he was to electrical 'atmospheres' did not consistently distinguish between flow and distance effects of his fluid, let alone have a conceptual-practical distinction enforceable upon his presumed paradigmatic followers (Home, 1972; Heilbron, 1979: 334-9, 376-7). More importantly, careful study of such post-Franklinian electricians as Beccaria, Aepinus, Coulomb and Cavendish fails to display a clear line of internally driven and programmed 'paradigm articulation' leading toward an instrumentalist quantification of electrostatics. Rather, one has a story of conceptual and institutional struggle, of contingencies and negotiations centring on the nature, function and significance of a variety of Leyden jars, condensers, air condensers, electrophori and Coulombian torsion balances (Heilbron, 1979: 373-489).[4] Where Kuhn packs the exemplary condenser into the paradigm from its beginning, Heilbron documents the battles to constitute and enforce any consensus about the basic conceptual issues and their concomitant instrumental manifestations. In any case, for a sub-set *only* of late eighteenth-century electricians, this struggle issued in a quantitative electrostatics, still requiring a more thorough mathematical formulation, by Poisson in 1811. Ironically, by that time such a culmination of the mathematicizing destiny proved to be temporary and unstable; for, whatever mathematical electrostatics was in the hands of Poisson, his proffered canonical version soon required revision and reinterpretation in the light of electro-chemical, and then electro-magnetic effects, hardwares and theories, hotly contested and promoted by their own overlapping sets of devotees.[5] In all this there is little sign of the Kuhnian paradigm with a destiny.

3. BACHELARD: THE GENESIS OF EXPERIMENTAL SCIENCES

Bachelard, like Kuhn, is concerned with the emergence of fields of mathematical-experimental science out of largely qualitative and discursive bodies of thought. These he terms 'pre-science', a generic term subsuming all non-scientific systems of thought addressed to nature or technics; *e.g.*, craft knowledges, common-sense rationalizations of phenomena, traditional medical theories, alchemy, magic, as well as all traditional systematic philosophies of nature from Aristotle to Priestley. All these pre-scientific discourses are seen as constituted by cognitive errors and pitfalls that must all be fully expunged before any scientific discourse can emerge. Some of these pathol gies include: appeal to analogy and metaphor drawn from any realm whatsoever; projection onto natural phenomena of fundamental affective structures and dispositions; naive empiricism, the grasping and generalization of one's first impression of the phenomenon; and classification and natural history that merely place phenomena within a set of conceptual pigeon holes imported from some other realm of discourse. Generally speaking, the conceptual frame is external to science itself, having psychological or cultural origins, and the phenomena are viewed passively through these frames (Bachelard, 1949; 1965; 1975). Any genuine science, according to Bachelard, is the antithesis of all this. Its concepts and indeed its objects are created from within itself. A science takes shape in some limited and strictly controlled technical-experimental context in which simultaneously, (a) phenomena are produced and systematically varied in their relevant parameters, and (b) a system of interdefining ('interfunctioning') mathematicized concepts is created which directs the material realization of the phenomena and the variation of their aspects. A science is not built from natural facts, nor deduced from a priori concepts; it subsists in the resonation or interplay of the couple (a) and (b), denoted by Bachelard respectively as 'technical materialism' and 'applied rationalism':

Technical Materialism <=> Applied Rationalism

Research in a constituted field of science is a dialogue of the two moments. The right side leads the way, by unfolding theoretical possibilities of variation and suggesting modes for their realization or objectification in the experimental hardware on the left (1949; 1975; Lecourt, 1977: 40-7, 60-70, 76-9).[6]

When a science comes into being, that is, when such a couple is initially constituted, an artificial technical realm comes into being in which phenomena are literally manufactured under the joint guidance of the system of mathematicized concepts and the instruments and experimental hardware in which those concepts have been realized. In an ironic jibe at positivist dogma, Bachelard terms any such realm of theoretically dominated artificial experience a 'phenomeno-technique'. By this he means that the phenomena of science are not discovered but made; not natural, but artificial. They are created and commanded in the light of theory and theory-loaded instruments and not by common sense, natural philosophical systems, or any other 'pre-scientific' mode of conceptualization (Bachelard, 1965: 60-1; 1975: 17; Gaukroger, 1976:

212-23).[7] It follows that any given science is born through a discontinuity, an 'epistemological rupture' with the relevant preceding pre-scientific discourse (Lecourt, 1977: 86; Gaukroger, 1976: 228).[8] The limited technical realm and the neat system of scientific concepts must be summoned into existence together; and the new-born couple lacks any relation to pre-scientific discourses.[9]

For Bachelard the Leyden jar is a pre-scientific conceptual monstrosity, created through pre-scientific imagination run amok. Electricity is a fluid; to store a fluid you use a jar -- so pump fluid from your generator into water contained in a glass jar (Bachelard, 1949: 148-50; 1965: 76). But, Bachelard counters, true electrical science would later conclude that electricity is not a fluid; that is not held in the water (something the eighteenth-century actors soon learned on a practical level); that the capacity of the jar for electricity relates not to its volume but to the inner surface area and, paradoxically to common sense, the thinness of the glass (short of being so thin as to rupture in discharge). There is a complete mismatch between the metaphors guiding the invention and practical use of the jar, and the true concepts later controlling its properly scientific use and construction. Electrical science begins not with the jar but with its transcendence: when it is shown that any object can be a capacitor which can realize the fundamental equations linking charge, potential, specific inductive capacity, surface area of plates and thickness of dielectric (Bachelard, 1949: 151). Such objects are scientific objects -- artifacts, not natural facts -- their construction and functioning commanded through the equations expressing the interfunctioning of the concepts. Novel conceptual possibilities, for example, equations giving the capacitance of sets of capacitors in series or parallel, equations which take counter-intuitive forms, can be foreseen and attempts can be made to realize them, hence enriching the concept of capacitance and developing the instrumental hardwares and their capabilities (1949: 154). Accordingly, Bachelard would date the science of electrostatics from the work of Coulomb, Cavendish and then Poisson; not from the pre-scientific fantasies of the Franklinists.

Bachelard's model of experimental science bears some relations of paternity to 'social constructivist' approaches to experiment. Later we shall develop these in articulating our notion of experimental natural philosophy. For the moment, however, we note some ways in which Bachelard's conception of the phenomeno- technique rests upon and legitimates a whiggish two-place historiography and is profoundly antithetical to the content and spirit of recent sociology of experiment.

Bachelard, like later sociologists of experiment, saw experimental hardwares as embodying conceptual structures, yet he saw them as 'frozen' conceptualizations whereas sociologists of experiment stress the fluidity and negotiability of meanings embodied in hardwares. By 'frozen' we mean, in the first place, that for Bachelard the development and articulation of a science can involve only minimal alteration of concepts once they have been acquired. Even the initial creation and embodiment of a set of science-making concepts constituted a permanent scientific achievement (1951: 25*ff*). In any given field of experimental science, the conceptual fabric initially constituted in the science-making rupture with pre-science cannot be undone by radically transforming, re-negotiating or rejecting the key concepts of the science. Although Bachelard saw mutually conditioning changes in the terms of the hardware-discourse couple as the content of serious research in mature fields, the conceptual changes he envisioned are of a moderate and mainly incremental kind: additions to the conceptual/hardware armoury, refinement or increasing precision on both sides of the couple, and the excision of the occasional gross error (Gaukroger, 1976: 227-34). 'Frozen' also means that Bachelard restricts the content of true sciences; only certain kinds of concepts can be frozen. For Bachelard the theoretical or 'applied rationalism' side of the hardware-theory couple is in fact always an 'interdefining' set of mathematicized concepts. Qualitative concepts or merely discursive theorizing plays no part. The birth of a science depends upon the creation (and phenomeno-technical embodiment) of such a system of mathematicized concepts bearing no relation to the previous pre-scientific discourse except that of negation. The distinction between pre-science and science is, on the conceptual side, a distinction between ordinary discourse and mathematics: only mathematics can be scientifically embodied in hardware, and the articulation of the theoretical side of a science is really the unfolding of new mathematical relations in terms of the existing set of mathematicized concepts. The relation 'science is to pre-science as mathematics is to discourse' underpins Bachelard's two-place historiography.

We can now begin to discern that Bachelard's enterprise falters in much the same way as Kuhn's. Their problems do not arise from their failure to agree on the site of the advent of scientificity, in the case of electrostatics or any other; but rather, from their attempts to locate such sites at all. As with Kuhn's search for first paradigms, Bachelard's regard for cataclysmically novel phenomeno-techniques is essentially whiggish in the most profound sense of the term, since it involves the search for some event or achievement upon which to pin the medal for 'finally-

before and indeed after the moment for the award of the prize.

In contrast we argue that here experimental natural philosophy is the relevant historical category. It was experimental natural philosophy that was undergoing change and transformation, its tortuous and contingent history issuing some times and some places in states and events to which whigs apply the epithets 'science', 'phenomeno-technique', or 'paradigm' in their scrambling to award epistemological merit points. The historiographies of Kuhn, Bachelard, and garden variety whigs beg for deployment of the category of experimental natural philosophy. The problem is how to accomplish this: ironically, both Kuhn and Bachelard provide hints as to how to go about displacing their own two-place historiographies. As we shall see in the next section, Kuhn tacitly admits to there being some sort of continuity of conception and practice across the pre-paradigm/paradigm divide. We shall suggest that Kuhn's intimations of continuity and process can and must be articulated through a conception of experimental natural philosophy lacking in his work. After we have begun in section 5 to define experimental natural philosophy, we shall find in section 6 that Bachelard's conception of the phenomeno-technique, suitably revised in constructivist terms, provides a key element for our model of the dynamics of experimental natural philosophy.

4. KUHN'S BRUSH WITH CONTINUITY

Despite the message conveyed by a first reading of *The Structure*, Kuhn never managed decisively to distinguish between pre-paradigm/paradigm stages in the development of an experimental field. But in his hesitation and ambivalence, Kuhn recognized the dynamics of something like what we would term experimental natural philosophy, a something he terms the 'Baconian sciences'. We will first examine his difficulties in *The Structure* and then scrutinize his conception of these Baconian sciences.

Kuhn's difficulties with the pre-paradigm/paradigm distinction appear in the opening portions of *The Structure*, and remain unresolved in his 1970 'Postscript'. Recall Kuhn's (1970: 14-5) description of the competing pre-paradigm schools of electricians, each characterized by its own metaphysical commitments and preferred domain of problems and phenomena. This makes the pre-paradigm schools seem very paradigm- or science-like, and it is a description holding for the pre-paradigm stages of other fields as well. For example, in discussing pre-paradigm

or science-like, and it is a description holding for the pre-paradigm stages of other fields as well. For example, in discussing pre-paradigm schools in physical optics, he (1970: 12-3) even observes that they each had different respective 'paradigmatic observations'. To critics he responded in the 'Postscript' in the spirit of his earlier ambivalent assertion that pre-paradigm researchers 'were scientists, [but produced] something less than science' (1970: 13), arguing that the transition to maturity of a science 'need not be associated with the first acquisition of a paradigm', for the members of the schools of the pre-paradigm period of a field 'share the sorts of elements which I have collectively labelled a paradigm' (1970: 179). What changes, Kuhn (1970: 179) now suggested, is not

> the presence of a paradigm but rather its nature. Only after the change is normal puzzle-solving research possible. Many of the attributes of a developed science which I have associated with the acquisition of a paradigm I would therefore now discuss as consequences of the acquisition of the sort of paradigm that identifies challenging puzzles, supplies clues to their solution, and guarantees that the truly clever practitioner will succeed.

This certainly seems a distinction without much of a difference, when we recall Kuhn's views on the effectiveness and maturity of the fluid school of electricians even before Franklin. Kuhn needed a rupture, a transition to maturity, but was left with weak indicators of its supposed occurrence: relatively greater consensus (the end of inter-school debate) apparently dependent upon relatively greater puzzle-defining and puzzle-solving power (which also instills confidence). But, given the paradigm-like virtues of the pre-paradigm schools, it is still not possible to see, in Kuhn's terms, why and how such points of superiority are made out and enforced upon the debating parties.

We conclude that Kuhn's difficulty in grounding his two-place historiography leads him in the direction of these vague intimations of 'continuity'. The underlying cause of his problem, we suggest, resides in his failure to recognize and deploy a category of natural philosophy. We shall now see how the absence of this category vitiates much of his understanding of the Scientific Revolution and permits (and demands) his use of the term 'Baconian sciences' to articulate the continuity that looms through his failure to enforce his own pre-paradigm/paradigm distinction.

As several of Kuhn's essays show, he sees the Scientific Revolution as having advanced on two loosely connected fronts (1977: 31-59, 136-7, 213-21). The first is the radical change in the pre-existing classical sciences, such as the mathematically oriented ones of geometrical astronomy, optics, mathematics itself and the study of motion, as well as in the bio-medical ones of medical theory/physiology and anatomy. The second is the initial development of some empirically and experimentally oriented Baconian sciences, the fields that, in *The Structure*, mature to their first paradigms at various stages in the eighteenth century. His schema largely ignores the great natural philosophical systems of the early modern period, the conflicts of which, we shall argue in Section V, defined the rhythm and moments in the Scientific Revolution. To Kuhn (1977: 53) they provide the 'metaphysical' elements or 'new intellectual ingredients' of the existing, shifting and nascent sciences in play on the two fronts.[11] He does not recognize the role of systematic natural philosophies *per se* in demarcating and conditioning the content of what he terms the sciences, both existing and nascent. Clearly, to speak of atomism as a 'new intellectual ingredient' bearing the metaphysics of the cluster of classical sciences, is not the same thing as to trace the construction and triumph of corpuscular-mechanical philosophies, their ongoing interaction with the classical sciences, and their relations with the emergent domain of 'Baconian' experimentation (Schuster, 1989)

Kuhn's (1977: 31-59) story of the Baconian sciences amounts to saying that a cluster of experimental 'sciences' [*sic*] emerged or were extracted from alchemy, magic and the crafts; were loosely fitted under the corpuscular-mechanical umbrella and imbued with the Baconian rhetoric about utility and progress; and then matured into true sciences. This is little more than an articulation of the model of the maturation of new eighteenth-century experimental sciences presented in *The Structure*, developing the narrative of the 'pre-paradigm' states of these sciences. But, because this story grounds the experimental *sciences* in their Baconian origins, it must veer away from the categories and two-place historiography of *The Structure*, and toward the hints of pre-paradigm/paradigm continuity present there. After all, Kuhn assumes that these sciences really existed from their initial seventeenth-century crystallization and that, in general, there were less mature scien es in the pre-paradigm stage, and more mature sciences in the post-paradigm stage. Kuhn's continuity, therefore, can only be a continuity of some kind of scientificity. Once again, he falls into difficulties by not having a working category of natural philosophy. Despite intimations of continuity and process, he fails to theorize natural philosophy as an historical category, and hence neglects to consider if the dynamics of natural philosophy might have had anything to do with these putative Baconian sciences and their long gestation to 'full maturity'.

5. NATURAL AND EXPERIMENTAL NATURAL PHILOSOPHY

The history of science in the early modern period, the age of the Scientific Revolution, and the eighteenth century is very much the history of natural philosophy, or more properly of competing systems of natural philosophy, and their relations with a set of more narrow, technical fields of inquiry. Some fields like astronomy or optics were mature, long constituted and undergoing significant alteration; others, like the experimental fields of the eighteenth century, were just emerging in the period. To miss this point is to forfeit any firm grasp on the historical trajectory of European science. Not only was the history of science made in the struggle for cognitive and social hegemony amongst the systems of natural philosophy but also the field of competing systems of natural philosophy constituted the matrix from, and ultimately against which, the experimental fields emerged (Schuster, 1989). It is necessary, first, to understand in generic terms what natural philosophy consisted in as a social and cognitive undertaking. This involves looking beyond the existence of numerous competing systems of natural philosophy in order to examine their general form, contents and aims as well as their mode of production and dissemination within a sub-culture. This will allow us to characterize experimental natural philosophy generically, and hence articulate its dynamics, whilst avoiding some explanatory pitfalls awaiting the historian of natural philosophy in general and the eighteenth-century problem in particular.

Every system of natural philosophy, whether of a generally Aristotelian, Neo-Platonic, mech nistic or Newtonian type, purported to describe and explain the entire universe and the relation of that universe to God. The enterprise also involved a concern with the place of human beings and human society in that universe. Each system of natural philosophy rested upon bour structural elements the contents and systematic relations of which went far toward defining the content of that particular system. These elements were: (1) a theory of substance (material and immaterial), concerning of what the cosmos consists, what kinds of bodies or entities it contains: (2) a cosmology, an account of the macroscopic organization of those bodies; (3) a theory of causation, an account of how and why change occurs; and (4) an epistemology and doctrine of method purporting to show how the discourses under (1), (2) and (3) were arrived at and/or how they could be justified, and how they constitute a system.

The existing specialist fields of science, those broadly delineated above in Kuhn's categorization of 'classical mathematical and bio-medical sciences', were pursued under the aegis of, and hence subject to contingencies within, the enterprise of natural philosophy. Each science was considered to be part of, or conditioned by, one or another system of natural philosophy. In the terms of Kuhn, and of Koyre whom he largely followed in this regard, the ' etaphysical' presuppositions of a science were supplied or enforced by a system of natural philosophy. The shape of a science, its direction of development, and indeed its very legitimacy often depended upon the character of its natural-philosophically enforced metaphysics, which itself might have been the outcome of conflict and debate. So, for example, the mechanical philosophy could be used either to marginalize scientific undertakings which were subsumed by and hence supported antagonistic systems of natural philosophy or to co-opt portions of otherwise 'dubious' scientific enterprises by reinterpreting them in terms of a mechanistic metaphysics, as was done with work of such non-mechanists as Harvey, Kepler, Gilbert and Bacon. Conversely, rapid change in a science and/or a shift in its social evaluation could lead to the constitution, alteration or abandonment of a natural philosophy through the borrowing or rejection of privileged images or models. Paracelsus' 'alchemization' of natural philosophy comes to mind here, as does the rise of the mechanical philosophy, which was grounded in the concerns of practical mathematicians and devotees of 'mechanics'.

Natural philosophies, however, were not mere collections of cognitive forms and aims. They were products of a sub-culture or, better, something like a Bourdieuian field of power and discourse relations (Bourdieu, 1975; 1981) in which the construction, modification and purveying of natural philosophies was a rich *sui generis* social enterprise, pursued in various institutional locations, with overlapping and partially competing bodies of discursive resources. There existed competing sub-species of natural philosophical discourse, all subject to complex and shifting pressures coming from, or issuing toward neighbouring fields of discourse, for example, in theology, politics, the practice and rhetoric of the practical arts, and the existing and nascent 'sciences'. To be a natural philosopher was to be engaged in the construction, interpretation and communication of the sorts of discursive entities we have described above. The individuals involved devoted themselves to natural philosophizing with serious and hard-won skills because of the value placed upon, and power conferred by, having the 'correct' view of nature. A correct natural philosophy would be expected to yield social and cognitive hegemony over the other players and over the institutional loci they occupied. It was also expected to enable the victors to determine and enforce their own particular notions about the proper content, order and goals of the existing and nascent 'sciences'. Finally, it was also expected to enable the victors to enforce their own views about the correct relations between natural philosophy and any and all 'contextual' discourses (and their institutional relations) of *perceived* relevance -- theological, political, cultural.

It is important to understand natural philosophy as a generic field of social and cognitive struggle, for it will allow us to define and explain the sense in which the advent of 'experimental natural philosophy' after 1650 was a mutation of the genus, a generic shift across the entire field of natural philosophy. We shall see that what Kuhn misleadingly describes as the birth of the Baconian sciences was actually a function of this change in the generic character of natural philosophizing, a change centring upon the discursive reconfiguration of notions of experiment and experience and of the relation of natural philosophy to the practical arts. What Kuhn calls the Baconian sciences, we shall see as experimental natural philosophy; where he sees those sciences continuously evolving over a century and a half -- albeit punctuated with the achievement of first paradigms -- we shall see the dynamics of the field of experimental natural philosophy, fostering the crystallization and demarcation of regional domains of experimental work and discourse. To explicate this shift, we require a brief account of aspects of the Scientific Revolution, centring on the field of natural philosophizing, the pattern of natural philosophical struggle, its key stages, and especially the struggle to define and enforce notions of experience and experiment, and of the role and function of the practical arts.

The Scientific Revolution may be viewed as a process of change and displacement within the field of competing systems of natural philosophy. The process involved the erosion and downfall of the dominant Aristotelian philosophy of nature and its replacement during the middle third of the seventeenth century by variants of the newly constructed mechanistic natural philosophy, which after a period of consolidation and institutionalization, were modified and partially displaced by the post-mechanist natural philosophies of Leibniz and especially Newton. The erosion of Aristotelianism in the sixteenth and early seventeenth century was connected with the proliferation of alternative natural philosophies of magical, alchemical and Hermetical colourations, and the mechanical philosophy was as much a response to the social, political and theological threats seemingly posed by these competitors as it was a response to Aristotelianism. The crucial moment in this process lies in the first two generations of the seventeenth century, an era of heightened conflict amongst Aristotelianism, its magical/alchemical challengers and nascent mechanism. Newtonianism was hardly the teleological goal of the process, but rather a complexly conditioned, contingent (and surprising) modification of the classical mechanism of the mid-seventeenth century.

Viewed this way, the Scientific Revolution takes on an interesting rhythm as a process of change and transformation of systematic natural philosophy (Schuster, 1989). There is a preliminary sixteenth-century stage, which will be termed the Scientific Renaissance, characterized by the erosion of Aristotelianism in some quarters, its further entrenchment in others, and by a ferment of revived alternative natural philosophies. There follows a 'critical' period (*ca.* 1590-1650) of natural philosophical conflict marked by the initial construction of mechanistic philosophies, and then a brief period of relative consensus on and institutionalization of a range of variants of the mechanical philosophy (*ca.* 1650-1690), punctuated and complicated by the advent of Newtonianism. It is during this post-1650 stage that we shall speak of a generic shift in the field of natural philosophy to 'experimental natural philosophy'. As we shall show, this shift was first instanced in the advent and acceptance of varieties of experimentally oriented, 'Baconized' mechanical philosophies but as a permanent generic shift across the field of natural philosophizing it persisted through the advent of Newtonian natural philosophy. Generic experimental natural philosophy, actualized in mechanistic and Newtonian species of natural philosophy, was the domain within which occurred the eighteenth-century formation of experimental fields. We shall argue that this occurred through processes largely having to do with the altered dynamics of the natural philosophical field, once it had become generically 'experimenal'.

The sixteenth century saw the recovery, assimilation and publication of a wide and confusing array of non-Aristotelian, 'alternative' natural philosophical systems: neo-Stoicism and Lucretian atomism; varieties of neo-Platonism, some more or less flavoured with Hermetic influences and variously amalgamated with alchemy, natural and demonic magic and cabala; and more eclectic and idiosyncratic ones derived from figures such as Telesio, Paracelsus, or even the radical Aristotelian Pompanazzi. One of the many factors powering this effusion of natural philosophical systems was a re-evaluation of the practical arts which first began to gain momentum amongst sections of the educated and natural philosophically literate elite in the sixteenth century. Important questions surrounded the social role of such practical arts as navigation, cartography, architecture and fortification, surgery, mining, metallurgy and other chemical crafts and the status of their methods, tools and results as 'knowledge'. These questions and the answers they received on the plane of natural philosophical discourse helped shape the aims and contents of competing systems of nature, some developments in the individual sciences, and, indeed helped prompt the emergence of generic 'experimental' natural philosophy (Rossi, 1970; Schuster, 1989).

The questions were promoted by the deeper social and economic structures of the 'long' sixteenth century, which extended from the late fifteenth to the early seventeenth century: by the historically high levels of population increase and price inflation; the expanding commercial capitalist economy; the enhancement of the power, and to some degree the size, of central governments; and the growth of the size, cost and technical complexity of military and naval hardware, logistics and administration. The number and diversity of potential patrons and clients of the practical arts increased. This, along with expanded literacy and access to print on the part of some master practitioners, their social allies and patrons, brought into being a social/cognitive field. There, practitioners competed for recognition and honour, whilst simultaneously contributing to the disparate chorus of claims that the practical arts and artisans deserved higher status, and that their skills and craft knowledge warranted the cultural status of 'science', in the contemporary sense of orderly discipline of intellectual worth (Rossi, 1970; Eisenstein, 1979: 520-74). The alternative philosophies of nature were important vehicles for those currents of opinion placing a premium on operative knowledge and on the search for command over the powers of nature. These systems could marshal powerful sentiments in favour of the combined practical and spiritual value of mathematics, and in favour of the natural philosophical relevance of the practical arts.

What we have termed the critical period of the Scientific Revolution (*ca.* 1590-1650) is characterized by a conjuncture between an unprecedented burst of conceptual transformation in the classical sciences -- instanced above in Kuhn's conception of the Scientific Revolution -- and a heightened, often desperate competition amongst systematic natural philosophies (some tied to utopian and irenic programmes of religious and social reform). The tendencies corrosive of Scholastic Aristotelianism -- the challenge of Paracelsianism, of Hermetically or alchemically tinged neo-Platonism, of calls for the re-evaluation of practical knowledge, of anti-Aristotelian rhetoric tied to the practice of the mathematical arts or classical mathematical sciences -- all took on greater virulence and urgency. From this emerged the mechanical philosophy which was self-consciously designed and sold by a handful of innovators in an effort to finesse and resolve (in their own favour) the natural philosophical crisis of the age (Schuster, 1989). All this occurred within a context of political, religious and intellectual turmoil which raised the stakes that actors perceived in the conflict of natural philosophies: stakes so perceived because of the presumed generic relations of natural philosophizing with political and religious issues and domains of discourse (Rabb, 1975; Popkin, 1979; Easlea, 1980). Since a genuine culture of natural philosophizing existed for the contemporary actors, problems and tensions in religion, politics and society were variously read and interpreted through the filter of natural philosophizing thereby entailing that these problems had some of their basis in the very confusion of competing systems.

The founders of mechanism hoped to resolve the conflict of natural philosophies in a way which was cognitively progressive, but religiously and politically conservative: by exploiting and co-opting recent achievements in the classical sciences, including the Copernican initiative, and by raising the premium placed upon mathematics and operative knowledge by sections of Renaissance opinion whilst avoiding the perceived religious, political and moral pitfalls of the alchemical, Paracelsian, Hermetic and eclectic atomistic systems then bidding to displace the Scholastic Aristotelianism (Lenoble, 1971; Easlea, 1980). Accordingly, the selection and moulding of discursive resources to form the mechanistic systems was a nice and dangerous task: first because it involved endorsing some values and aims characteristic of magical-alchemical systems, whilst explicitly opposing them as such; second, because the resulting product was itself intended to displace Aristotelianism in the institutional centres of natural philosophy. Mathematics was construed to avoid any hint of neo-Platonic mathematical fancies. Yet, as in neo-Platonism, mathematics was to be the very language of nature. Experience was identified with experiment, itself rhetorically modeled upon the disassembly and reassembly of machines, so as to marginalize alchemical and Paracelsian accounts of experience as an affect-laden, spiritually sanctioned and uplifting intuition of otherwise hidden relations and correspondences. Yet, as in magic and Paracelsianism, operative command over nature was sought through an active experiential engagement with nature. The mechanical philosophy was also constructed so as to embody an arguably orthodox 'voluntarist' vision of God's relation to nature and to mankind, so as to avoid collapsing the divine into nature and/or elevating man, as seeker of operative knowledge as well as wisdom, to the level of a 'magus', a status morally and cognitively unacceptable to mainstream Catholic and Protestant thought alike.

Francis Bacon, whose writings would in the third stage of the Scientific Revolution be melded with varieties of mechanism to form the first species of experimental natural philosophy, also needs to be placed in the context of the critical stage. He was a brilliant 'bricoleur' of disparate sixteenth-century values and a filter and refiner of the catch-cries, polemics and attitudinal shifts characteristic of the Renaissance stage. He addressed at the level of natural philosophical culture the debates over the status of practical knowledge, the aims and methods of 'useful' education of gentlemen, and the Protestant stress on cultivating socially useful, this-worldly vocations. Like the mechanists he ntended the simultaneous displacement of Aristotelianism and its magical, neo-Platonic and alchemical challengers. Of particular importance was his co-option and re-working of the issue of practical knowledge and the practical arts (Rossi, 1970). He proposed through his method discourse to capture the labour and results of practical artisans, whilst denying their potential status as natural philosophers (Martin, 1988). Through this tactic and his method discourse in general, he recast the notions of experiment and experience over against Scholastic conceptions on the one hand, and Paracelsian, magical and alchemical ones on the other. He rhetorically constructed a category of sober, public, replicable, and communicable experience, to be instantiated mainly in experiment and in the findings of the practical arts, suitably winnowed and sorted. This methodically defined experience was, of course, aimed as much for control and command over nature as for explanation and understanding of it. In the succeeding generation, all this was open to selective adoption and reinterpretation by triumphant mechanists.

The third stage in the process of the Scientific Revolution is characterized by the dissemination and acceptance of varieties of the mechanical philosophy, and by the melding of mechanism's own discourse of mathematics and experiment with a rhetoric of method and experiment loosely attributable to Bacon which emphasized experimental grounding, tentative theorizing, exploitation of instruments and possible technological benefits. A consensus formed around an experiment-oriented corpuscular-mechanical natural philosophy [hereafter ECM]. It was a loose consensus, to be sure, for ECM was not one definitive system, but rather a rough template from which were fashioned specific variants. The consensus was nevertheless real, especially when compared with the conflict of natural philosophies in the two earlier stages (Schuster, 1989).[12]

ECM's Baconized rhetoric of experiment hardly sufficed to guide the articulation of the natural philosophy or the sciences now subordinated to it (Schuster, 1989). Yet, like similarly ineffective and vague method doctrines, it helped shape the way knowledge claims were assembled, negotiated and entrenched or rejected (Schuster & Yeo, 1986; Richards & Schuster, 1989). It also functioned at the institutional level in providing rhetorical resources for solidifying and delimiting legitimate practitioners and practices, as in the apologetical and programmatic rhetoric of the Royal Society (Wood, 1980; Hunter, 1981; Shapin & Schaffer, 1985; Dear, 1985; Golinski, 1989). In these two related guises the method rhetoric of ECM helped constitute the sense in which natural philosophy had become 'experimental', at least in an initial 'declaratory' sense.

The crucial point is that because consensus over ECM had replaced previous fragmentation and dissensus, the advent of ECM constituted a shift in the generic field of natural philosophy. In addition, because ECM drew upon the resources of Baconianism and early mechanism to reconfigure the *declared* meaning and place of experience and experiment in natural philosophical discourse, the field of natural philosophizing as a whole now begins to deserve the generic title of experimental natural philosophy. This characterization holds across the later advent and relative triumph of varieties of Newtonian natural philosophy, including Newton's inductivist method rhetoric. We can therefore see that Kuhn's description of the opening phase in the maturation of the Baconian sciences applies to this very situation but that for any talk about Baconian sciences we must substitute the foregoing conception of the newly triumphant natural philosophy of ECM. That is, the continuity in the development of experimental fields that Kuhn discerned belongs not to 'Baconian sciences', but to a shifted generic field of natural philosophizing.

6. A SOCIAL CONSTRUCTIVIST VERSION OF THE PHENOMENO-TECHNIQUE

We must now begin to unpack the dynamics of experimental natural philosophy. The central problem is: What did it mean in a natural philosophical context to do experiments, and to be concerned with instruments and experimental hardwares, their production and their accounted outcomes? If we lose our way here, our reconstruction of Kuhn and Bachelard's problematic will have been in vain. Interestingly, Bachelard comes to our rescue, provided he is viewed through spectacles provided by the constructivist sociology of scientific knowledge. The key to our problem resides in Bachelard's conception of a phenomeno-technique, the coupling of experimental hardware and a specific theory or discourse. Two generations before the advent of a sociology of experiment, Bachelard saw that the phenomena or objects of inquiry of experimental sciences are manufactured in and by experimental hardwares, because such hardwares are themselves developed, used and understood as embodying conceptual structures. The conception of the phenomeno-technique is Bachelard's major bequest to the emerging sociology of experiment and new historiography of experimental sciences. Through it he began to suggest the constructed rather than given character of scientific 'facts', since they literally are the product of hardware-discourse couples, and, strictly speaking, exist only in relation to and in the vicinity of such couples.[13]

Construed as Bachelard literally intended, the phenomeno-technique holds little promise for our problem. Indeed, it is the cornerstone of his two-place historiography; as we saw in section III, it trades upon his distinction between qualitative discourse which cannot be phenomeno-technically embodied and mathematics which can. The notion of the phenomeno-technique can, however, be rendered more historical and therefore more useful, by refracting it through recent findings in the sociology of scientific knowledge. A revised conception of the phenomeno-technique can then be transplanted into a conception of experimental natural philosophy so as to suggest a basis for understanding the dynamics of the latter and the processes of crystallization of fields from within it. There is in Bachelard no concession to the interpretative flexibility of conceptual and practical resources commonly granted in the recent empirical literature on the sociology of scientific knowledge, especially constructivist sociology of experiment. Accordingly, there is no sense in Bachelard that hardware-discourse couples, no matter how apparently stable, are contingent products, outcomes of social processes of closure, and always subject to re-negotiation of the elements on either side of a couple. We must conserve Bachelard's fundamental recognition that hardwares are embodied theory and that the meanings of their 'outputs' or performances -- the nature of the 'objects' or phenomena they manifest -- are functions of that same embodiment of theory. We must, however, correct Bachelard's rigid insistence that embodied concepts can only be mathematical, permanent and non-contingent.

The seminal work on experiment of such sociologists of scientific knowledge as Collins (1985), Pinch (1986a) and Pickering (1984) points precisely in the direction needed. Collins pioneered the attempt to show that decisions about how to make and use experimental hardwares and about the meanings of their outputs are social rather than natural or methodological accomplishments. His work on debates about the existence of gravity waves -- and supposed gravity wave detectors -- prompted his well-known conclusion that debate 'about what counts as a competently performed experiment...is coextensive with the debate about what the proper outcome of the experiment is' (Collins, 1985: 128). Hence, closure of a debate about the meaning of competence constitutes the 'discovery' or 'non-discovery' of a phenomenon and the determination of its properties, for the time being. We see the same interpretative fluidity and correlative need to study the social modes of closure of debate about what an instrument is and what 'objects' it discloses, in Pinch's empirical studies and in his conception of the 'externality of observation' (Pinch, 1986a; 1986b).

Broadly speaking, the goal of the social processes of closure, or specification of relevant parameters, is to achieve a relatively unproblematic understanding of the experimental hardware so that its resulting 'output' can then be accounted to be saying something relatively unambiguous about the nature of the objects of inquiry. This would apply to any historical-sociological characterization of Franklin's attempt to impose his theory of the Leyden jar, and hence his version of the jar and of the phenomena or objects -- such as his electrical fluid -- it discloses. When and where the Leyden jar, that 'monstrosity' for Bachelard, was agreed to be what Franklin said it was, and to be doing what Franklin said it was doing, there and then Franklin's version of the 'electrical fluid' existed. Franklin's version of electrical fluid existed in a certain way of using and glossing this hardware. As ways of glossing -- the discourse of electricity -- changed, so did the objects manifested by this hardware, which was also thereby changed because somewhat differently glossed and managed.

Sociological analyses of experiment suggest that Bachelard's conception of the phenomeno-technique needs to be 'loosened up'. We require a Bachelard with a fluid and reinterpretable hardware-discourse couple, a Bachelard who would acquiesce in the following premises of a 'social constructivist' Bachelardian historiography. First, instrumental and experimental hardwares are accountably 'adequate' materializations of theories and concepts agreed/enforced as 'adequate' at some point in the past and 'until further notice'. Second, the performance and output of a hardware is doubly discursive and socially negotiated because of the character of the hardware as consensually frozen discourse and because any output requires further interpretation, evaluation and deployment in its intended argumentative contexts. Third, the existence and continued use of a hardware does not show that the discourse accountably manifested in it is 'true to reality'. It merely shows that in the past discursive resources were used to devise hardware and standards of its 'adequate' performance and output, and that its performance is still judged adequate to goals and interests themselves open to revision and contestation. Fourth, adequate command over a hardware does not depend upon possessing a good scientific theory of that hardware. The history of the practical arts demonstrates this, and so does the history of experimental natural philosophy. In sum it would seem, contra-Bachelard and pro-Collins *et al.*, that phenomeno-techniques are fluid and socially constructed, not given in the canon of true or permanent science.

Something, nevertheless, is still missing from our conception of the situation and tactics of early experimentalists. In these early struggles to construct and impose hardware-discourse couplings, what was at stake were embodiments within hardwares of *qualitative* conceptions. One was not dealing with 'inscription devices', those highly standardized hardwares which 'black-box' many layers of theory and are used in large arrays in modern experimental science. The study of inscription devices, as originally conceived by Latour and Woolgar (1979: 45-53, 63-9), emphasizes scientists' literary construction of 'facts' through their mobilizing, interpreting and deploying in further 'texts' the already textual outputs of highly black-boxed devices. Early experimental science or natural philosophy was not characterized by such inscription devices, but by debates and struggles over the initial black-boxing of qualitative theories and concepts into non-inscription hardwares.[14] This entails that we must identify the sources, sites and dynamics of the discursive materials being so embodied. The historical fact of the matter is that many of these discursive resources that Franklin and his contemporaries were struggling to mould to, and embody in various hardwares were of natural philosophical provenance. Natural philosophy was the paramount discursive medium from which was drawn the conceptual stuff for embodiment, and to which such embodiments, if successfully imposed, were supposed to speak: as applications, confirmations, explications or justifications of natural philosophical claims. The dynamics of experimental natural philosophy involved, therefore, attempts to constitute and establish embodiments of natural philosophical discourse within experimental hardwares.

Kuhn's pre-paradigm/paradigm distinction and Bachelard's pre-science/science distinction really mark moments in larger and longer processes of negotiation and enforcement of particular hardware-discourse couplings. These were carried out under the perceived social and cognitive hegemony of natural philosophical discourses that had already become generically 'experimental'. Indeed, below we shall argue that the tendency of experimental natural philosophy to fragment into what Kuhn and Bachelard would like to see as new experimental sciences arose from the largely unintended consequences of this dynamic of seeking couplings between hardwares and natural philosophically relevant discourse. But, before tackling the wider question of the nature and dynamics of experimental natural philosophy, we begin with the case of the Leyden jar and Franklin, looking at it as an example of a moment or node in this sort of process.

7. FRANKLIN AND THE LEYDEN JAR: EMBODIMENT OF NATURAL PHILOSOPHICAL DISCOURSE

The Leyden jar was an artifact of which the design, construction and use was based upon well established consensual understandings of concepts drawn from natural philosophical discourse and from the reservoir of contemporary common sense discourse. The coneption and construction of the jar hardware was, of course, dependent upon ordinary language understandings of terms like 'jar', 'pin', and 'water', terms which might acquire extra technical 'spins' as they were associated with the design, use and agreed 'output' of the newly stabilized artifact. More important from our standpoint, in order to build a jar in the first place, some stability and consensus had to exist about terms like 'electrical effluvia' or 'fluid'; about 'electrics' and 'non-electrics'; and about properties of the flow of fluid or effluvia and the sparking, seepage and possible containment of the same. The understandings of these objects and their properties or relations were already variously embodied in a collection of 'electrical' hardware-discourse couplings: electrostatic generators and other canonical pieces of apparatus for producing and manipulating electrical effluvia or fluid. In other words, there existed a domain of identifiably 'electrical' hardwares and the discourse they embodied. The embodied discourse about objects and properties, most importantly electrical effluvia or fluid, owed its origin and its ultimate point of reference and significance to the realm of natural philosophical discourse, with its agreed intellectual/social hegemony over ways of approaching, understanding and commanding nature. But, the borrowed elements of natural philosophical discourse were already acquiring 'spin' and specificity by their use and embodiment in this electrical realm. Terms like 'electrical effluvia' were supposed to be able to be reintegrated into overarching discourses about the system of nature but their meanings were increasingly the product of the contingencies of struggle and negotiation within the more narrow, but growing domain of electrical hardwares and embodied discourses.

Within the nascent electrical domain, there also existed routine rules of procedure, rules of thumb for working and manipulating hardwares. One such rule was derived from Dufay's work on conductors and non-conductors: in modern terms, to charge a conductor, the conductor has to be insulated; in their terms, to transfer fluid or effluvia to a non-electric, that non-electric has to be supported by an electric. As we mentioned above in Section 2, the jar's apparent violation of this rule of procedure provided one of the leads for Franklin's attempt to reinterpret what the jar was, what it did, and hence what were the relevant properties of the electrical fluid.

Franklin viewed the jar as a self-discharging device rather than a tool for delivering dramatically large jolts through discharge chains, although he did that as well. This concentration on the phenomenon of self-discharge to neutrality was the key to shaping his discourse of the jar and thus helps explain the densities and gaps in that discourse. His coming to 'clarity' about the jar occurred roughly as follows. After discharge the jar could be shown to be in a neutral condition. Now, electrical fluid had certainly entered the water (later the glass) from the generator via the prime conductor and pin. In discharge the only possibility was that the surcharge of fluid had travelled back up the pin and around the circuit to the outer surface, which must therefore have been in an under-charged condition. That in turn helps explain why the jar had to be grounded to be charged, in apparent contravention of Dufay's rule, for when the inside of the jar is charged somehow an equal amount of fluid is driven off, repelled, into the ground from the outside of the jar. If there is no grounding, there is no effective repelling of the fluid and hence no effect.

Viewing the situation this way allows us to conclude firmly that Bachelard is wrong to denigrate Franklin's understanding and construal of the jar as monstrous and pre-scientific. We have seen that the jar in intention and performance represented a reasonable artifactual embodiment of accepted natural philosophical and common sense understandings. It 'worked'; but only on condition that one arguably violated Dufay's rule, another apparently stable element in the emerging electrical domain and one which itself summarized a range of stabilized couplings in that domain. In Franklin's own grappling with the Leyden jar, everything proceeded reasonably from the standpoint of the culture of experimental natural philosophy as the available discourse and hardware were reinterpreted and renegotiated into a new coupling, proffered for wider acceptance. In his accounts of the jar, he revised available discourse about electrical fluid, its flow, conservation and self-repulsion across the glass in a way 'adequate' in his and others' view to allow manipulation of the jar, explanation of its salient effects, and improved understanding of the rationale of its design and construction (Cohen, 1956: 452-7; Heilbron, 1979: 330-1; Franklin, 1774: 180-2).

Nevertheless, all of this does not signify that Franklin had achieved some sort of paradigmatic, science-making breakthrough in the senses conveyed by Kuhn's two-place historiography. We can see this very clearly by recalling the nub of Kuhn's understanding of the Franklinian paradigm, or exemplary problem solution. As noted earlier, Kuhn claims that the key to Franklin's understanding of the jar was his possession of a concept of electrostatic induction, paradigmatically embodied in his explanation of the jar and making possible the critical distinction between the conductive phenomena of the fluid and its exertion of repulsive force at a distance (upon itself). This distinction allowed Franklinist electricians to pass beyond the Leyden jar itself and to conceive of and manipulate instances of what we termed the 'condenser as such'. Therefore, this distinction ultimately permitted the emergence of a fully mature, *i.e.*, mathematical, electrostatics.

What did Franklin 'know' about this modern conception of 'induction'? Strictly speaking only this: his newly revised discourse of the jar had a moment in it which we recognize as the place where *we* would insert the term 'induction' in all its modern connotations. Franklin simply spoke of repulsion across the glass (Franklin, 1774: 229) or of fluid 'driven out' of the outer surface when the inner surface is charged (1774: 180; Heilbron, 1979: 331). The momentum of his account drove him in this direction: his awareness of the Newtonian possibility of speaking about repulsion gave him the natural philosophical warrant and some materials for so doing. But, Franklin did not thematize this repulsion as a central theoretical problem or topic. He simply inserted this 'repulsion' into his ongoing talk as a plausible move in his universe of discourse, given what he did thematize--the equivalence of input and output of fluid. In further rounds of negotiation of this and other hardware-discourse couples, Franklin never consistently seized upon, thematized or deployed the distinction between fluid flow and its distance effects. At best, the distinction lurks in his discourse. He had a fitful, unfocused recognition of it, for example, when he improved upon Canton's discourse of Canton's own 'induction' experiments, as they are termed in standard histories of science (Franklin, 1774: 302-6; Cohen, 1956: 524-31; Heilbron, 1979: 376-7, 336-9). The main thrust of Franklin's subsequent articulation of his discourse was away from focusing on the distinction and toward a discourse of the dynamics of 'electrical atmospheres' (Home, 1972). As we argued earlier, the construction and deployment of the distinction was a long, contingent struggle played out over two generations amongst the very electricians whom Kuhn saw as working within a Franklinian paradigm-with-a-destiny.

Kuhn's conception, we reiterate, is not tenable. We do not therefore say in whiggish fashion that Franklin missed or clouded the distinction; nor, in the style of Heilbron (1979: 335-6) do we accuse Franklin of 'bad physics' and try to salvage some progressive 'phenomenological core' of his achievement hidden within his increasing concern for atmospheres and their dynamics. Franklin had good reasons to proceed in the directions he did. In the felicitous terminology of Barnes (1982: 30, 45-53), Franklin and the Franklinists had to negotiate the next applications of concepts such as 'repulsion across the glass', and the 'flow/distance effects distinction' because there is no algorithm for doing this, no prescriptive paradigm with a destiny. Hence nobody was being unscientific, pre-scientific or downright stupid for not negotiating in the direction of those 'applications' that would be approved of by Kuhn, Heilbron and other implicit or explicit whigs.

Franklin's hardware-discourse couple was neither a pre-scientific monstrosity without a future nor an exemplary problem solution constituting a paradigm with an algorithmically derivable destiny. It was neither a science-making achievement nor a cognitively pathological failure to constitute a science. Franklin constructed and attempted to enforce upon his peers a revised natural philosophical discourse about electrical fluid, its flow, conservation and self-repulsion, a revised discourse adequate to explaining and manipulating the jar which in turn came to be seen by him and his allies as embodying that discourse. He did not have classical electrostatics, or even its concept of induction; the application of his revised discourse of electricity to new or revised hardwares was not obvious or algorithmic. His was a relatively successful bid to establish and stabilize a particular couple in an expanding and contested region of experimental natural philosophy. His couple is precisely the sort we expect to see coming and going under the regime of experimental natural philosophy. Its subsequent history under that regime, and its role -- as resource and as object of renegotiation -- in the development of the electrical domain within experimental natural philosophy, consists in a concatenation of socially negotiated contingencies like the ones that first constituted it.

In other words, Franklin's proffered couple was a node in a process consisting in a struggle to constitute, stabilize and institutionalize such couples, and, if possible interlink them with additional couples to which they were accountably 'related'. Experimental natural philosophy was the generic field of discursive resources, social sites and power relations in which this process took place. The relative stabilization of a couple, until further notice, constituted a node. On the one hand, such stabilization always involved some shift, reinterpretation or renegotiation of existing objects of natural philosophical discourse -- their nature, properties and relations -- and their embodiment in new or reconsidered hardware. On the other hand, any stabilized couple provided a resource, a point of reference, for possible flexible use in further rounds of struggle to constitute new or revised couples. A node, therefore, was a relatively settled couple, in which stabilization (for the time being) involved some drift or shift in existing natural philosophical understandings and which was available for deployment in a fluid and reinterpretable way in further rounds of attempts at similar stabilizations.

All this suggests an analogy that we shall develop further in later studies. The process of constructing and negotiating couples under the aegis of a supervening reservoir of natural philosophical discourse may be likened to the process of making particular, contextually shaped utterances within a given natural language. The proffered couple resembles the utterance, and the hegemonic natural philosophical discourse resembles the background language system which both shapes and makes possible particular utterances, and yet which exists largely in virtue of them and their unfolding, contingent history. On this analogy, Franklin, like any other experimental natural philosopher, was, so to speak, conversing about the hardware -- with himself and with his fellow natural philosophers. His language, his reservoir of resources for talking and the rules for doing so were supplied by his knowledge of natural philosophy with its already existing electrical domain and, of course, by the resources of ordinary language and usage. But, as conversations are always in context, so Franklin's hardware-discourse coupling for the jar was one such utterance and, like an utterance in natural language, it might or might not have become a resource, topic or source of models for further rounds of utterance (Watchirs & Schuster, 1987).

8. DYNAMICS OF EXPERIMENTAL NATURAL PHILOSOPHY

With the example of Franklin in place and armed with our sketch of the concepts of hardware-discourse couples, nodes, and an electrical domain, we proceed to a fuller articulation of our general conception of experimental natural philosophy and its dynamics. We have introduced the conception of an emergent electrical domain within the field of experimental natural philosophy. A general understanding of such domains and their formation is central to our model of experimental natural philosophy. It should already be clear that the idea of phenomeno-technical nodes must be viewed in relation to it. We now proceed to articulate the idea of emergent domains, making use of our heuristic analogy between the dynamics of language and experimental natural philosophy. Domain formation fundamentally arose from the imperative to understand and command hardwares through natural philosophical discourse, an imperative built into generic experimental natural philosophy. After all, to have discoursed about hardwares in, say, traditional craft languages would not have been part of natural philosophy' at all, and could not have triggered the dynamics we are about to explore.

Consider first the simple case of an isolated hardware-discourse coupling which has settled down as a 'fact', the hardware taking on, for the time being, a settled meaning. This would mean that the hardware's construction, use and output are established and it would also mean that the phenomena or objects realized or manifested by the hardware, so understood, also exist. The very settling down of the couple means that there has been produced a particular utterance, a particular version of the natural philosophical discursive elements in question. This process would be compounded by the local, technical 'spin' given to those elements of ordinary language deployed in that coupling. The discourse of the hardware in this coupling would thus be an amalgam of mutually conditioning natural philosophical and ordinary language terms.

In our simple case the utterance would have little significance for domain formation. It would likely either be ignored at the level of grand natural philosophical systematics or be trivially reabsorbed into one or another system as a small fact, illustration or piece of evidence. The situation would be different, however, if there were a set of accountably 'related' couples and if several actors or groups of actors were involved in the process of 'settling' and relating them. This would then constitute a nascent domain. Specific couples within the domain could become resources and objects of contention among and within schools of natural philosophizing, as competition over priority and hegemony within the domain begins to emerge as an issue. In later rounds of debate, understandings of some couples might be taken for granted by some or all of the actors, as they mobilize them in their respective endeavours to define and impose new couples. This in turn would tend further to crystallize the domain and give it a density throug a shared repertoire of couples and understandings.

Domain formation may be likened to the formation of a dialect, if we focus upon the subtle meaning-shifts of natural philosophical terms and their amalgamation with terms derived from ordinary language. That is, a domain is like a dialect within an overarching 'official' language of natural philosophy. The existence of such an emergent regional natural philosophical discourse, a dialect, is tolerably clear in the case of electricity on the eve of the advent of the Leyden jar. As we saw above, 'conduction', 'generation', 'sparking' and 'leakage' were all examples of ordinary language terms increasingly given technical spins and connotations in the emergent domain of electrical work prior to the advent of the jar. These had merged with such more properly natural philosophical terms as 'fluid', 'effluvia' and 'electric' to form an electrical dialect within natural philosophy. The dialect itself was used in projecting and constructing the jar. Elements within the dialect were shared by such actors as Franklin, Nollet and Watson who then contended definitively to construe the jar. The dialect was in turn modified through the enforcement of Franklin's discourse of the jar as definitive, for the time being.

In a nascent domain, and in and around its growing cluster of couples, there will develop low-level rules, maxims of procedure, state-of-the-art expectations about entities in the domain and their uses, relations and behaviours. This is symptomatic of domain and dialect formation. Typically, such rules, principles and maxims have no partisan natural philosophical significance. That is, such 'domain rules and principles' do not serve as resources for one school of natural philosophy over against others, because they are neutral amongst the contending schools, being easily communicated and/or translated amongst them, and because they are agreed to be necessary for anybody's functioning within the domain. We have already instanced Dufay's rule for charging conductors, a rule of thumb that needed to be followed and could be understood by electricians regardless of their natural philosophical predilections. Indeed, Dufay's rule was a working maxim within a larger discourse on 'electrics' and 'non-electrics', a classification of their instances and properties. This entire classificatory discourse might be considered to have been a set of domain rules in that it did not uniquely presuppose or support one natural philosophical ontology as opposed to any other and yet was necessary for navigation within the electrical domain. The initial development and acceptance of Gray's work on conduction in rather ontologically neutral terms also indicates the presence of a set of domain rules (Cohen, 1956: 370; Heilbron, 1979: 245-9). No particular species of natural philosophy had a special purchase on the discourse of conduction; it was intelligible and necessary for anyone playing the electrical game. One can go so far as to say that the emergent electrical dialect consisted precisely in such domain rules, principles and maxims. Their ontological neutrality (or easy translation) bespeaks the crystallization of a domain as well as the tendency for natural philosophical systematics and conflict to be displaced by engagement within the domain.

It is important to see that domains and dialects would have formed above and beyond the conscious intentions of any natural philosophers to demarcate and dominate them. The mid-seventeenth-century shift in the generic character of natural philosophy set in train the struggle to attach natural philosophical discourse to hardwares, to speak about hardwares in the name of something as important as natual philosophy. When an experimental natural philosopher tried to enrol his peers in a bid to embody natural philosophical discourse in a couple, he was speaking in the name of his favoured species of natural philosophy. Peers might or might not follow such a bid, but the issue was engaged at the level of natural philosophizing, meaning that for all concerned there were at stake the existence and properties of the fundamental furnitures of nature and the laws of their behaviour. Nevertheless, although the main intention may have been to extend the empires of particular systematizations of natural philosophy, domain and dialect formation were set in train as unintended consequences. Conversely, as these processes continued and domains and dialects began to crystallize, natural philosophers could then thematize these tendencies and began to contend for supremacy in demarcating and dominating the nascent domains, thereby accelerating the process.

Our conception of domain and dialect formation casts further light on Kuhn's quandaries over the pre-paradigm 'schools'. These were defined primarily in terms of metaphysical predilections or, in our terms, natural-philosophical commitments. As we saw, Kuhn had trouble factoring such schools into his model. On the one hand, he had to depict some of the schools as too 'paradigm-like'; on the other, to introduce into the schools a suitable degree of 'Baconian' fact gathering so that they would remain on the pre-paradigm side of the historical divide. The phenomenon of domain and dialect formation cuts across the terrain marked by such Kuhnian schools even when we construe them in our terms as divergent and variously organized strands of natural philosophical opinion. When Kuhn sought science-making breakthroughs in one or another school he was looking for the wrong thing in the wrong place. His alternative strategy of tracing the long-term maturation of the 'Baconian sciences' also missed the mark for he misconceived experimental natural philosophy and its peculiar dynamics.

The conceptions of domains and dialects and of their dynamics constitute the kernel of our model of experimental natural philosophy, the key to our reconfiguring of Kuhn's and Bachelard's problematic. In order to make further headway with this model, we must turn to the issue of the earliest stage of experimental natural philosophy, grafting our notions of domains and dialects onto it. Experimental natural philosophy was the terminal phase in the history of the generic field of natural philosophy, during which practitioners were concerned to assert the hegemony of natural philosophical discourse over existing and yet-to-be constructed hardware-discourse couples. For example, it declared its relation to, and hegemony over, much of the existing field of the practical arts, hoping thereby to replace or displace localized, parochial craftmen's discourses with natural philosophical discourse.

At first, in the hands of Bacon and the early mechanists, experimental natural philosophy was merely a 'declaratory' doctrine. It asserted its in-principle connection to, and right of command over, existing constellations of couples in the practical arts and those couples already co-opted into competing Paracelsian or natural magical species of natural philosophy (Bennet, 1986). Initially, it had little to show for this effort in terms of actual co-option and reconfiguration of the discourses of hardwares. When Bacon, the early mechanists or the triumphant advocates of ECM addressed the practical arts and artisans, it was initially in this imperative and declaratory fashion, the cash value of which was, employing Pinch's terminology (1986b), to invoke a new, global 'evidential context '-- natural philosophy -- for the use, evaluation and understanding of the hardwares of the practical arts. Moreover, in so far as Bacon, the early mechanists and the advocates of ECM also sought to defeat and displace competing systems of natural philosophy, systems that had already co-opted portions of the practical arts, the exponents of experimental natural philosophy had either to reject such co-opted portions, or redescribe them in terms of their own favoured Baconian/mechanist discourse.

In broad terms this is the very process by which the generic field of natural philosophy became 'experimental'. As outlined in Section 5, it required Bacon and the mechanists first to reshape the terms experience and experiment in ways different from and 'superior to' their use in Aristotelianism and species of magical natural philosophy, and second to claim hegemony over the practical arts in the name of their revised natural philosophical discourses. With the wide triumph of ECM, this declaratory form of experimental natural philosophy came to define the field of natural philosophizing. It instigated a struggle to co-opt and link hardwares, and that process led, largely unintentionally, toward domain and dialect formation. This is not to say that the declaratory style of experimental natural philosophy could and did produce dramatic and quickly accelerating results in the form of constellations of hardware-discourse couples and a correlative crystallization of domains. Rather, the existence of the declaratory moment of experimental natural philosophy was critical in both starting and shaping the type of natural philosophy that led over the next century and a half to domain formation and the terminal fragmentation of the traditional field of natural philosophy. For, without declaratory experimental natural philosophy, those domains eventually called experimental sciences would not have developed. The evolving domains of couples in the practical arts and technics would not have been declared objects of natural philosophical attention. These domains instead would have continued to evolve apart from the elite culture of natural philosophizing.

One would therefore expect that early in the history of experimental natural philosophy, the main thrust was to cover and co-opt hardwares with discourse aimed at supporting one's own natural philosophy and attacking others. Technical details of hardware and outputs may have been relatively less interesting in themselves than the drive to enrol the hardware and its newly insinuated discourse for systematic purposes: to display the correct natural philosophical system, method, concepts or values. For example, the politics of inter-system debate fuelled the early controversies about the meaning, nature and function of the barometer and the air pump, with each side viewing its own discourse of the hardware as manifestation of, or evidential support for, their preferred natural philosophical objects and relations (Middleton, 1964; Shapin & Schaffer, 1985). Nevertheless, once a hardware became the focus of such inter-system competition, pressures appeared toward the formation of a domain, a dialect and domain rules. For there to have been debate and struggle over accounting a hardware and its output, there must have been some overlap in the glosses and accounts rendered, competing but partially overlapping accounts becoming resources for further rounds of debate about this hardware and other accountably 'related' ones. If the use of a given system of natural philosophy to provide discourses of hardwares produced a regional dialect, then the pressure of inter-system conflict on the level of natural philosophy would only accelerate and inflame the process of domain formation.

The early development of the domain of electricity displays some of these features. Prior to Gilbert the electrical domain, if one may so stretch the term, consisted in the electrostatic properties of amber, effects grasped in ordinary language as well as at the margins of natural philosophical discourses of natural magic and curiosities. Having to hand only the amber effect along with his own lengthened list of similarly 'electric' substances, Gilbert (1600) inscribed the electrical domain within natural philosophy primarily as the 'non-magnetic'. This was in keeping with his goal of d stinguishing magnetic phenomena and their 'spiritual' causes from electrical phenomena and their merely mechanical, effluvial causes. His understanding of the magnet, of the constitution and 'output' of the magnetic compass, and of his magnet experiments in general, were relative to and largely dictated by his concern to establish a novel system of 'magnetic' neo-Platonic natural philosophy that embodied and supported a modified Tychonic cosmology. He claimed hegemony over such craft practices as navigation, mining, and metallurgy, and a right to co-opt discursively findings and facts embedded in them. In fact, he did co-opt and re-gloss existing couples, most notably on the critical issue of the compass.

Gilbert's natural philosophizing of the magnet, its phenomena and chief hardware, was too important and impressive a gambit in the field to be ignored by his competitors. So, for example, the efforts of early mechanists such as Descartes (1644) were directed at re-glossing Gil ert's experimental work in mechanistic terms, rather than at extending the number and type of magnetic couples or articulating their technical specifications, matters that tended to remain in the ambit of master artisans. Descartes also devoted considerable attention to preserving and capturing the 'cosmic' significance of magnetism, the keynote of Gilbert's system. He replaced Gilbert's story of the cosmos making and binding role of the spiritual magnetic force with a mechanist's story of an equally cosmic magnetism which was now the purely mechanical effect of a species of corpuscle of particular and peculiar shape and size moving in suitably configured aggregations of ordinary 'third matter'.

In view of this titanic natural philosophical struggle to co-opt and 'enrol' magnets and magnetism, in Latour's (1987) slightly bizarre terminology, the phenomenon of electricity remained insignificant. Despite Gilbert's multiplication of the category of 'electrics', it was trivially mechanistically explained, and still occupied a marginal place in the hierarchy of natural philosophical topics. In the seething cauldron of seventeenth-century natural philosophical conflict, it was a speck of flotsam. The non-existence of a range of 'electrical' phenomena and the provision of the known effects through the most trivial of 'hardwares' -- the substances themselves -- guaranteed this marginal status. This helps explain the flatness, for example, of Boyle's perfunctory attempt (1675) to co-opt electricity to his brand of corpuscular mechanism. As is well known, this situation only began to change in the early eighteenth century. The electrical domain began to widen and thicken, with heightening natural philosophical interest, only with Hauksbee's multiplication of electrical phenomena and constitution of prototypical electrostatic generators, together with the implication in Newtonian natural philosophy that electrical phenomena might bespeak a significant power in nature. The subsequent multiplication of hardwares and effects both raised the natural philosophical stakes and promoted those phenomena of domain, dialect and domain rule formation discussed above.

9. HISTORIOGRAPHICAL TRIALS OF NATURAL PHILOSOPHY

We have argued that a conception of experimental natural philosophy and its dynamics can reorient our view of the Kuhn/ Bachelard problematic, as well as our first order picture of the trajectory of experimental science in the latter stages of the Scientific Revolution and in the eighteenth century. Our model invokes a conception of generic natural philosophy as a field of social and cognitive struggle and views experimental natural philosophy as a consequence of a shift within that field. Experimental natural philosophy was expressed initially as a declaratory doctrine, which set in train a largely unintended dynamics of domain and dialect formation, through the imperative placed upon the co-opting and commanding of hardware-discourse couples by means of natural philosophically relevant resources. The resulting struggle to establish and stabilize couples bespeaking one or another species of natural philosophy fostered that drift of discourse and clustering of 'related' hardwares that we have termed domain and dialect formation. As regional domains and dialects developed, so too did wider self-identifying communities of practitioners, working in and through their respective domains. At certain times and places actors who had developed skills and reputations within a domain could in turn seize upon it as something worth contending over thereby fomenting more 'proto-disciplinary' struggles to demarcate, define and command particular domains. It is these latter formations and types of struggles that have prompted from actors and historians alike discourses on the 'birth' of this or that field of experimental science.

The implications of this model are clear for any two-place historiography whether of whiggish-inductivist, Bachelardian or Kuhnian temper. We submit that the task of historians of experimental sciences in this period is not to seek the mythic *origins* of sciences. We do not need tales of the sudden discovery of proper methods, metaphysics or models, which supposedly shattered the debilitating fascination exercised by putatively non- or pre-scientific beliefs and enterprises. Nor do we require myths of the slow gestation of happily 'pre-formed' sciences. Rater, the historian of early experimental science must first grasp the social and cognitive nature of natural philosophy, both generically and in its historical manifestations as competing systems, and then describe the dynamics of change, development and ramification by which were produced those moments, turns and crystallizations misread as *de novo* outbreaks or inventions of experimental 'sciences'. We need to know not why and how the experimental fields count as sciences, but what natural philosophy and experimental natural philosophy were, and how and why they ramified and ultimately dissolved into the experimental fields. In short, we need an historiography of latent processes and of unexpected and unintended endings, not an historiography of heroic births, whether miraculously sudden or long gestated (Butterfield, 1965: 43, 51-2, 59-60).

This is not to assert, however, that our model provides a sufficient historical description or explanation. It is preliminary, suggestive and mainly heuristic in intent. This becomes clear if one analytically distinguishes what we might term primary and secondary processes shaping the historical trajectory of experimental natural philosophy. Our model is of the primary process only, and yet the primary process was of course only ever played out in and through specific sites, struggles, and contingencies of a social, institutional and biographical nature that are the stuff of detailed state-of-the-art social history of science and historical sociology of scientific knowledge. Theorizing about the primary process does not entail that we conjure out of existence these contingencies and conditions. It means instead that we place the detailed analysis of local conflicts and contingencies in a new context, emphasizing that in and through those local contingent events and struggles there played upon the actors discursive and social imperatives, resources, opportunities and constraints bound up in the existence of a culture of experimental natural philosophizing. Just as one cannot deduce the details of any instances of secondary process from the model of primary process, so one cannot fully grasp events in the secondary process without understanding the actors moving in and through the field of experimental natural philosophy. One must view actors' machinations as partly enabled and constrained by the field and conversely one must view their machinations as at each turn instancing it, (re)constituting it, altering it and driving on its dynamics.

Understood in this manner, our model prompts three broad sets of historiographical observations about experimental science and natural philosophy. First, it follows from our model that the historiography of experimental natural philosophy cannot simply consist of snapshots of particular contrversies, debates and closures no matter how exquisitely executed these pictures may be. Even Shapin and Schaffer's (1985) brilliant reconstruction of Boyle and Hobbes' debate over the status, meaning and output of the air pump does not capture some generative core or founding step in experimental natural philosophy. Instead, it reconstructs a moment in the early phases of experimental natural philosophy, characterized by continuing debate over systematic hegemony and also by the need to specify and enforce there and then some socially constructed and stabilized protocols for obeying, in this concrete situation and hopefully in others, the imperative to form couples impregnated with natural philosophical discourse. Shapin and Schaffer document a local struggle to constitute what we may term a Latourian social-technical actor network of Anglican divines, gentleman amateurs, and skilled, committed experimental corpuscular-mechanists. Certainly, the newly emergent field of declaratory experimental natural philosophy did not dictate or determine the details of this local passage of struggle. Yet, it was a medium of resources and constraints, of imperatives and possibilities, in which took place this struggle, in so far as it was perceived as a struggle in and over the philosophy of nature, as that enterprise and field of conflict was lived and breathed, not only in London and Cambridge, but also in Paris and Pavia. Moreover, subsequent rounds of doing experimental natural philosophy in other times and places did not in principle or in fact have to occur in or through this locally constituted network. Anglican divines and English latitudinarian gentlemen were not always and everywhere the essential players in other locally woven social-technical-discursive nets, and neither did the particular practices and protocols constructed and enforced in their net have any stability or transferability except in so far as they registered in the repertoire of resources possibly available to others -- elsewhere and at other times -- concerned to play (versions of) the game of experimental natural philosophy. As we do not comprehend an utterance apart from its language, so the language, as a structure undergoing contingent processes of change, does not really exist outside the ongoing concatenation of utterances. So, beyond and within the snapshots, depictions of secondary process, there must be correlative attention to the field of experimental natural philosophy, to the web of imperatives, opportunities and constraints it wove in and through each contingent situation. This web's texture must be taken to have altered over time, as it played and was played through concatenations of contingent events and encounters, from a declaratory moment to nascent and then thickening domains to self-conscious struggles over domains in preference to struggles about overarching systematizations.

Second, just as snapshots will not suffice for recasting the history of experimental natural philosophy, so hasty and total contextualizations of episodes occlude our understanding of the historical processes involved in the trajectory of experimental natural philosophy. For example, Shapin (1980; 1981) and Schaffer and Shapin (1985) advance a broad socio-political contextualization of natural philosophy, asserting connections between natural philosophers' struggles to assert and enforce conceptions of the natural order and the social forms and protocols of experimentation, on the one hand, and struggles to preserve or reform conceptions of the larger social order, on the other hand. As against this, our model would imply that the field dynamics of natural philosophy and experimental natural philosophy cut across and through any such immediate larger contextual determinations, when they did exist. Moreover, it would suggest th t concern with relations between the political order of natural philosophizing and the larger political order was neither essential to the dynamics of natural philosophy, nor a necessary or sufficient cause of its existence and ongoing social reproduction. Any such concerns of particular actors were contingent, secondary factors, certainly conditioning and refracting how they played couplings in their favoured domains, but certainly not the complete set of constraints, resources and aims available to and pressing upon them.

Third, our model gives some orientation to the present proliferation in the literature of confusing aperçu of 'natural philosophy' caused by the awakening of interest in the category of natural philosophy on the part of some historians of eighteenth-century science. Heilbron (1980), for example, is eager to package the developments we have grasped as 'domain/dialect formation' under the rubric of 'experimental physics', an activity he can easily see developing toward the emergence of relatively autonomous fields of mathematico-experimental science. His sober professors, academicians and scientists, practitioners of experimental physics, strive in instrumentalist terms toward the maturation of the mathematical/experimental fields, properly eschewing unhelpful, obstructive questions about 'matter theory'. But what, then, have they to do with the dramaturgical shenanigans of the sort of natural philosophers discovered by Schaffer (1983), maestros of morally and politically edifying public experimental spectacles? How, moreover, do we relate the activities of either group to yet a third body of more utilitarian and pedagogically motivated popular natural philosophical lecturers, the texts of whose courses of natural philosophy are organized around apparently independent chapters on such areas as mechanics, optics, hydrostatics, astronomy, electricity, chemistry and other topics (Cantor, 1982)? Should this proliferation and heterogeneity drive us to downplay the historical validity of the category 'natural philosophy', in the style of Heilbron, or, should we seek to define some metaphysical, methodological or even discursive consensus that really united all parties? We suggest our model does permit a glimpse of the wood comprised by these trees, without the expedients of instrumentalist denial or idealist desperate measures. Heilbron's proto-physicists were, of course, operating in particular nooks and crannies of eighteenth-century natural philosophical space, the academies, societies and universities being their favoured haunts. Hence they tended to seek the definition and control of their favoured experimental domains in and through the resources and rituals worked out in and for such sites. But, it remains the case that Heibron's proto-physicists were playing the game of hardware-discourse couplings within and through nascent domains and dialects; that these were shared with denizens of other sites; and that they accordingly were in fact natural philosophers doing experimental natural philosophy. The secondary process imperatives of their particular sites and styles may have increasingly incited them to define themselves over against competitors as the only legitimate inhabitants of the emerging domains; but this does not excuse them from being numbered by historians amongst the members of eighteenth-century experimental natural philosophy, and it certainly does not mean that they alone should engage our understanding because they partook of some privileged destiny of science-making.

Analogous observations hold for Schaffer's dramaturgical agitators. Their hardware-discourse couplings, their readings of hardware construction, uses and outputs, had different accents and interests attached and they were carried out in different sites for different audiences under different rituals and protocols. Moreover, they of course carried out their work in their own corner of natural philosophical space and they too tended to arrogate to themselves the very definition of natural philosophy in the name of the meanings they had invested in particular couplings before their favoured intended audiences. But, to the degree that they too played the hardware-discourse coupling game, they too were natural philosophers, although not the only ones.

As for natural philosophical courses and textbooks increasingly organized around disparate regions of phenomena, problems and their attendant clusterings of hardwares, they do not bespeak the non-existence of natural philosophy, nor should they cause puzzlement when juxtaposed with the Heilbronian or Schafferian conceptions of natural philosophy. Rather these books and courses are symptoms of that creeping, partly unintended, but increasingly intentionally exploited fragmentation of natural philosophy, marked by the shift from concerns over systematic hegemony to those over definition, demarcation and dominance of domains.

We return to the notion of natural philosophy as a social-cognitive field in which a struggle for hegemony was pursued in various institutional settings on the basis of partially overlapping and competing bodies of discursive resources. As we have argued, this was just as much true in the mid-eighteenth century as it had been in the mid-seventeenth, provided one allows for the relative homogenization of discursive resources ushered in by the advent of ECM and then Newtonianism, and provided that one also factors in the dynamics set in train by the advent of the experimental imperative across the field; that is, the accelerating tendency toward fragmentation of the field, arising from the phenomenon of domain formation, which was fostered by attempts at hardware-discourse couplings wherever they occurred--in the lecture room, public lectures and dramatic displays, the sittings of academies and societies, or the private cabinets of learned amateurs. The existence of heterogeneous types, styles and settings of experimental natural philosophy is perfectly comprehensible on this basis. There was no consensus in natural philosophy in the sense of a dominant style or site of experiment, ontological commitment, or characteristic attention to discourses of religion and politics. There was, at the level of primary process, only the field and its dynamics played in and through local sites and struggles. In sum, then, the advent of a constructivist sociology of experiment challenges us to re-write the history of experimental science in a style and temper unknown to Kuhn and Bachelard, not to mention garden variety whigs; but that enterprise can only be pursued through the formulation and deployment of revised categories of natural philosophy and experimental natural philosophy. This undertaking will immunize our efforts against the dual pathologies of merely micro analysis and hasty, totalizing contextualization, and it will affect our understanding not only of the eighteenth century, but also, as we have hinted, of the process of the Scientific Revolution as well.

ACKNOWLEDGEMENTS

We would like to thank Homer Le Grand and Rod Home for comments on a draft of this paper. Versions of our argument were presented to meetings of the Australasian Association for the History, Philosophy and Social Studies of Science and the Society for the Social Study of Science, and we profited from the discussions on those occasions. Research for this paper was supported in part by grants from the University of Wollongong, Board of Research and Post-Graduate Studies, and the Commonwealth of Australia Post-Graduate Scholarship scheme.

NOTES

1. Kuhn's use of the term 'gradual' does not undermine this judgment, because the overall effect of his discussion is to program the future outcome of much subsequent history into the moment of assimilation. The denouement of 'articulation' in mathematization is given as a destiny from that point.
2. The tendency for Kuhnian internalist historiography to rejoin traditional whiggism is discussed in Watchirs (1983).
3. Kuhn (1970: 21): 'Ever since prehistoric antiquity one field of study after another has crossed the divide between what the historian might call its prehistory as a science and its history proper. These transitions to maturity have seldom been so sudden or so unequivocal as my necessarily schematic discussion may have implied. But neither have they been historically gradual, coextensive, that is to say, with the entire development of the fields within which they occurred. Writers on electricity during the first four decades of the eighteenth century possessed far more information about electrical phenomena than had their sixteenth-century predecessors. During the half-century after 1740, few new sorts of electrical phenomena were added to their lists. Nevertheless, in important respects, the electrical writings of Cavendish, Coulomb, and Volta in the last third of the eighteenth century seem further removed from those of Gray, Du Fay and even Franklin than are the writings of these early eighteenth-century electrical discoverers from those of the sixteenth century.' Kuhn's relation to Koyre in this regard is discussed in Watchirs (1983: 8-15).
4. This, of course, is a reading of Heilbron much conditioned by commitment to a social constructivist position on facts, testing and experiment. Heilbron himself leans in quite a different direction, back toward the point where orthodox Kuhnianism of paradigm-as-destiny encounters good, old fashioned whiggery. For, above and beyond the mass of

detail in Heilbron's study; that is, whenever the forest of contingent complexity is glossed for the supposed historical big picture, the literary effect of positivist, instrumentalist progress toward quantified electrostatics is produced. All struggle, debate and indeed stupidity aside, mature electrostatics of textbook form was going to eventuate, once Franklin had done his work (Heilbron, 1979: 336, 377-8, 389, 396, 415, 436-8).

5. The reader can take this remark as a promissory note for a more social constructivist historiography of early nineteenth-century electricity and magnetism that is yet to be written, but which can be spied with the proper set of spectacles.

6. To paraphrase Bachelard (1975: 61): the meaning of a concept must include the technical conditions of its material realization.

7. On Bachelard's view, therefore, the various sciences are each sui generis; each has its own specific system of concepts and related instrumental armoury. No single, transferable, general scientific method can explain the genesis of such sciences or their respective contents and dynamics. Indeed, Bachelard produced a typology of philosophical images of science -- conventionalism, positivism, empiricism -- by assessing the degree to which they each tend toward an idealist or materialist misconstrual of the interaction of concepts and instruments which constitutes fields of science. (Bachelard, 1949: 4-5; 1975: 139; Lecourt, 1977: 60-1).

8. Needless to say there is a problem here. If there is a sharp break and displacement of discourses, in what sense, for Bachelard, was there a relevant preceding pre-scientific discourse in that area, domain, or field? Our revision of Bachelard in the direction of the sociology of scientific knowledge and through the mobilization of the concept of experimental natural philosophy overcomes this difficulty.

9. Pre-scientific discourses, for Bachelard, are culturally, historically and psychologically contingent discourses which mistakenly try to make contact with some 'real' given world. By treating objects as external givens, pre-scientific discourses create obstacles to their own development: there is no hope of creating possibilities of variation within a discourse immobilized by its own delusion of reflecting reality. What Bachelard's account cannot therefore accommodate is the fact that the sciences traditionally have operated under methodological and epistemological glosses making precisely the same sort of realist claims. This raises the post-Bachelardian question of what functions such glosses do possess in the life of those fields Bachelard would denote as sciences (Schuster & Yeo, 1986: xix-xx; Richards & Schuster, 1989).

10. Koyre and Kuhn taught us that whiggism in the history of science means a presentist perspective, a judging of the past by the transient standards of the present. But, if we go back to Butterfield (1965) himself, we find a richer, more complex sense of whiggism. It consists first of all in the creating of hypostatized historical categories which embody what we think we value in the present. These hypostatized categories are then set to work, combatting and overcoming equally hypostatized forces of negativity, so that gradually or suddenly, but always dramatically, our valued category emerges full-blown. Kuhnian and Bachelardian two-place historiography is therefore whiggish in the most serious professional historian's sense of the term. See Section 9.

11. This is the 'Koyréan' Kuhn speaking, first recognizing the often dominant role of deep ('metaphysical') preconceptions in the practice of any mature science and then going beyond Koyre in conceding that metaphysical commitments other than the 'Platonic' can underlay successful, normal scientific traditions.

12. It is fashionable, and correct, to maintain that late seventeenth mechanists, at least in England, made considerable capital out of what often was a necessity -- political, religious or technical -- to acknowledge 'spirit' and active agency in nature (Henry, 1986). It is not correct then to infer that the 'mechanical' philosophy is a misplaced or non-

existent historical category. One simply needs to distinguish within the textual body of any particular mechanist's system a level or dimension of declaratory or self-glossing discourse, where the system purports to gloss itself--sum itself up for presentation to audiences. Such glossing need not have been consistent with, or even an 'adequate' representation of, the nuts and bolts of the system of which it is a part. At this level most mechanical philosophies will declare themselves opposed to magic, alchemy and the claims of 'spirit', quite independently of where and when some concessions need to be made elsewhere in the discourse. Cartesianism is a nice negative example. Completely mechanist and anti-spirit at the declaratory level, Cartesianism harbours that curious 'first matter' which Descartes insists behaves fully mechanically, but which easily might be taken to have non-mechanical capabilities. In an English context, a hypothetical 'Descartes' himself might have seen fit to have pointed this out, whilst still insisting at the declaratory level on the fully mechanical character of his natural philosophy.

13. Moreover, Bachelard's model of science had two other major implications for the history and social studies of science that have only recently begun to be realized: (1) that all set piece method discourses are necessarily ineffective in the terms they proclaim for themselves (Schuster & Yeo, 1986; Richards & Schuster, 1989); and (2) that if science is 'phenomeno-technical', its analogies to and relations with technology may be richer, more subtle and nuanced than existing models of science and technology suppose (Latour, 1987).

14. It is interesting to place these observations in the context of Pinch's (1986b) illuminating analysis of 'externality of observation' in highly black boxed modern instrumental complexes. Experimental natural philosophers were often in a position analogous to those who, in Pinch's account of the 'solar neutrino telescope', would only be willing or able to acknowledge observation of a 'splodge on a graph'; that is, the first 'thing' in the negotiable chain of possible 'things' made observable by the device. Having no hardwares black-boxed, early experimental natural philosophers were attempting to make manifest in their hardwares immediately perceivable 'objects' or properties of objects. The objects and properties were discursively realized and embodied in natural philosophical talk, and correspondingly the hardwares were problematical in themselves as objects--what they did and what they manifested (as said and thought in natural philosophical terms). Another interesting contrast between studying modern experiment and experimental natural philosophy is that even Pinch, who attends closely to levels of hardware embodiment (and debates about them) and to the historical development of such embodiments, is actually more concerned to analyze levels of *argument* than levels of *hardware embodied meaning*.

REFERENCES

Bachelard, G. (1949). *Le rationalisme applique*. Paris: Presses Universitaires de France.

Bachelard, G. (1951). *L'Activite Rationaliste de la physique contemporaine*. Paris: Presses Universitaires de France.

Bachelard, G. (1965). *La formation de l'esprit scientifique*. 9e ed., Paris: J.Vrin.

Bachelard, G. (1975). *Le nouvel esprit scientifique*. 13e ed., Paris: Presses Universitaires de France.

Barnes, B. (1982). *T.S.Kuhn and social science*. London: MacMillan.

Bennett, J.A. (1986). The mechanics' philosophy and the mechanical philosophy. *History of science*, **24**, 1-28.

Bourdieu, P. (1975). The specificity of the scientific field and the social conditions of the progress of reason. *Social science information*, **14**, 19-47.

Bourdieu, P. (1981). Men and machines. In *Advances in social theory and methodology: Toward an integration of micro and macro sociologies*, ed. K. Knorr-Cetina & A.V. Cicourel, pp. 304-17. London: Routledge Keegan Paul.

Boyle, R. (1675). *Experiments and notes about the mechanical origin or production of electricity*. London. Old Ashmolean Reprints 7, Oxford, 1917.

Butterfield, H. (1965). *The whig interpretation of history*. New York: Norton.

Cantor, G. (1982). The eighteenth century problem. *History of science*, **20**, 44-63.

Cohen, I.B. (1956). *Franklin and Newton: An inquiry into speculative Newtonian experimental science and Franklin's work in electricity as an example thereof*. Philadelphia: American Philosophical Society.

Collins, H.M. (1985). *Changing order: Replication and induction in scientific practice*. London: Sage.

Dear, P. (1985). *Totius in Verba*: Rhetoric and authority in the early Royal Society. *Isis*, **76**, 145-61.

Descartes, R. (1644). *Principles of Philosophy*. V.R. & R.P.Miller (trans), Dordrecht: Reidel (1983).

Easlea, B. (1980). *Witch-hunting, magic and the new philosophy: An introduction to the debates of the scientific revolution*. Brighton, Sussex: Harvester Press.

Eisenstein, E. (1979). *The printing press as an agent of change: Communication and cultural transformation in early modern Europe*. Vol 2, Cambridge: Cambridge University Press.

Franklin, B. (1774). *Experiments and observations on electricity*. London. I.B.Cohen, ed., Cambridge: Harvard University Press (1941).

Gaukroger, S. (1976). Bachelard and the problem of epistemological analysis. *Studies in the history and philosophy of science*, **7**, 189-244.

Gilbert, W. (1600). *De magnete*. London. Rept., trans. P.F. Motteley, New York: Dover (1958).

Golinski J.V. (1989). A noble spectacle: Phosphorus and the public culture of science in the early Royal Society. *Isis*, **80**, 11-39.

Henry, J. (1986). Occult qualities and the experimental philosophy: Active principles in pre-Newtonian matter theory. *History of science*, **24**, 335-81.

Heilbron, J. (1979). *Electricity in the seventeenth and eighteenth centuries*. Berkeley: University of California Press.

Heilbron, J. (1980). Experimental natural philosophy. In *The ferment of knowledge: Studies in the historiography of eighteenth century science*, ed. G.S. Rousseau & R. Porter, pp. 357-87. Cambridge: Cambridge University Press.

Home, R. (1972). Franklin's electrical atmospheres. *British journal for the history of science*, **6**, 131-51.

Hunter, M. (1981). *Science and society in restoration England*. Cambridge: Cambridge University Press.

Kuhn, T.S. (1963). The function of dogma in scientific research. In *Scientific Change*, ed. A.C. Crombie, pp. 347-69. London: Heineman

Kuhn, T.S. (1970). *The structure of scientific revolutions*. 2nd ed., Chicago: University of Chicago Press.

Kuhn, T.S. (1977). *The essential tension: Selected studies in scientific tradition and change*. Chicago: University of Chicago Press.

Latour, B. (1987). *Science in action: How to follow scientists and engineers through society*. Milton Keynes: Open University Press.

Latour B. & Woolgar, S. (1979). *Laboratory Life: The social construction of scientific facts*. London: Sage.

Lecourt, D. (1977). *Marxism and epistemology: Bachelard, Canguilhem, Foucault*. Trans B. Brewster, London: New Left Books.

Lenoble, R. (1971). *Mersenne ou la naissance du mécanisme*. 2e ed, Paris: Vrin.

Martin, J. (1988). Knowledge is power: Francis Bacon, the state and the reform of natural philosophy. Ph.D. dissertation, University of Cambridge.

Middleton, W.E.K. (1964). *The history of the barometer*. Baltimore: Johns Hopkins University Press.

Pickering A. (1984). The hunting of the quark. *Isis*, **72**, 216-36.

Pinch T. (1986a). *Confronting nature: The sociology of solar neutrino detection*. Dordrecht: Reidel.

Pinch T. (1986b). Toward an analysis of scientific observation: The externality and evidential significance of observation reports in physics. *Social studies of science*, **15**, 3-36.

Poisson, S.D. (1811). Sur la distributionde l'electricité a la surface du corps conducteurs. *Memoires*, Académie des sciences, Paris, 1-92 & 163-274.

Popkin, R.H. (1979). *The history of scepticism from Erasmus to Spinoza*. Berkeley: University of California Press.

Richards, E. & Schuster, J.A. (1989). The myth of feminine method: A challenge to gender studies and social studies of science. *Social studies of science*, **19**.

Rossi, P. (1970). *Philosophy, technology and the arts in early modern Europe*. New York: Harper & Row.

Rabb, T.K. (1975). *The struggle for stability in early modern Europe*. New York: Oxford University Press.

Schaffer, S. (1980). Natural philosophy. In *The ferment of knowledge: Studies in the historiography of eighteenth century science*, ed. G.S. Rousseau & R. Porter, pp. 55-91. Cambridge, Cambridge University Press.

Schaffer, S. (1983). Natural philosophy and public spectacle in the eighteenth century. *History of science*, **21**, 1-43.

Schaffer, S. (1986). Scientificdiscoveries and the end of natural philosophy.*Social studies of science*, **16**, 387-420.

Schuster, J.A., (1989). The scientific revolution. In *The companion to the history of modern science*, ed. G. Cantor *et al.* London: Routledge, Chapman and Hall.

Schuster, J.A. & Yeo, R., eds. (1986). *The politics and rhetoric of scientific method: Historical studies*. Dordrecht: D. Reidel.

Shapin, S. (1980). The social uses of science. In *The ferment of knowledge: Studies in the historiography of eighteenth century science*, ed. G.S. Rousseau & R. Porter, pp. 93-139. Cambridge: Cambridge University Press.

Shapin, S. (1981). Of gods and kings: Natural philosophy and the politics of the Leibniz-Clarke dispute. *Isis*, **72**, 187-215.

Shapin, S. (1984). Pump and circumstance: Robert Boyle's literary technology. *Social studies of science*, **14**, 481-520.

Shapin, S. & Schaffer, S. (1985). *Leviathan and the air pump: Hobbes, Boyle and the experimental life*. Princeton: Princeton University Press.

Watchirs, G. (1983). The maturation of physico-mathematical fields in Kuhnian historiography: The case of eighteenth century electrostatics. BA honours dissertation, University of Wollongong.

Watchirs, G. (1989). Theories of scientific discourse: Beyond objectivity and subjectivity in the historiography of science. Ph.D dissertation, University of Wollongong.

Watchirs, G. & Schuster, J.A. (1987). The eighteenth century problem: Natural philosophy, experiment and discourse in post-Kuhnian perspective. Annual Conference of the Society for the Social Study of Science, Worcester, Mass.

Wilde, C.B. (1982). Matter and spirit as natural symbols in eighteenth century natural philosophy. *British journal for the history of science*, **15**, 99-131.

Wood, P. (1980). Methodology and apologetics: Thomas Sprats' *History of the Royal Society*. *British journal for the history of science*, **13**, 1-26.

University of Wollongong

BRUNO LATOUR

THE FORCE AND THE REASON OF EXPERIMENT

For Françoise Bastide. In memoriam

1. ARCHIMEDES ON THE BEACH

Archimedes, who was a kinsman and friend of King Hiero, wrote to him that with any given force it was possible to move any given weight; and emboldened, as we are told, by the strength of his demonstration, he declared that, if there were another Earth, and he could go to it, he could move this one. Hiero was astonished and begged him to put his proposition into execution, and show him some great weight moved by a slight force. Archimedes therefore fixed upon a three masted merchantman of the royal fleet, which had been dragged ashore by the great labours of many men, and after putting on board many passengers and the customary freight, he seated himself at a distance from her, and without any great effort, but quietly setting in motion with his hand a system of compound pulleys, drew her towards him smoothly and evenly, as though she were gliding through the water. Amazed at this, then, and comprehending the power of his art (*sunnoesas tes tecnes ten dunamin*), the King persuaded Archimedes to prepare for him offensive and defensive engines to be used in every kind of siege warfare. (Plutarch, 1961: xiv, 78-9)

In this famous report of what may be the oldest public scientific experiment, several features are remarkable. The performance has almost certainly never been staged (at least with a fully loaded ship);[1] it is thus a *tale* of a staged thought experiment, but a story which for hundreds of years played a continuous role in shaping the relations between Kings, mathematics, war and mechanics. It is a public show before all the assembled 'media'. It is a direct application of a theoretical demonstration that Archimedes had just completed following a Platonist research program that Plutarch sketches in the paragraph before. It is Archimedes himself who takes the initiative of boasting to the King that he can move the Earth; the King, quite reasonably, challenges him to a 'show down' by way of a smaller scale public experiment before believing in the demonstration, as if he was unable to be convinced by the strength of mathematics alone; but it is the King who, in an instant, makes the connection of this striking but futile experiment with a technical and military research program headed by Archimedes (*o demiurgos*) to protect Syracuse against the Romans.

H. E. Le Grand (ed.), Experimental Inquiries, 49–80.

But, why was Archimedes so emboldened by his apodeictic demonstration that he could move the planet and why was Hiero so amazed by the smaller scale show, impressed (as the text tells us) by the power of technique? Because 'some great weight may be moved by a slight force'. Is it possible to reverse the age old balance of forces? Can one man become stronger than the multitude? And this, not just ideally in a thought experiment about going to another planet and moving the Earth, but in practice, on the beach. Where the labour of many sailors and slaves had been necessary to move the ship ashore, the arm of an old, dignified free man was sufficient. A multitude becomes less strong than one, provided you add to one's inner workings a system of compound pulleys. Is this not quite sufficient to appeal to Hiero? A king is never physically superior to the multitude he commands. Archimedes, the physicist, is physically superior to the multitude because he knows how to balance proportions over a fulcrum. This reversal of force is of enormous importance for Hiero and since then for every Prince seduced by a Star Wars program.

According to Plutarch (xvii, 4-5), however, Archimedes himself is much more struck by another reversal:

> And yet, Archimedes [after equipping Syracuse with war machines] possessed such a lofty spirit, so profound a soul, and such a wealth of scientific theory, that although his inventions had won for him a name and fame for superhuman (*ouk anthropines*) sagacity, he would not consent to leave behind him any treatise on this subject, but regarding the work of an engineer and every art that ministers to the needs of life as ignoble and vulgar, he devoted his earnest efforts only to those studies the subtlety and charm of which are not affected by the claims of necessity.

Back to fundamental research. After a little detour through wizardry, demiurgy and weaponry -- like Szilard, Einstein and so many others since -- Archimedes goes back to geometry, to this extraordinary Planet War in which apodeictic reasoning can *reverse* the multitude of objections that common sense, the senses, tradition and normal decency bring against it. To be sure, he accepts Hiero's challenge, but this staged experiment is nothing compared to being able to *demonstrate* that any one weight could lift any other, no matter how big or to prove *against* all contrary evidences that he could move the Earth. This is what, in a Platonist world, really counts.

In this famous text which marks the start of Science, Technology and Society studies, there are four reversals of forces, each compounding the other. First, he, Archimedes, in a somewhat paranoid thought-experiment, could move the Earth provided he had a lever and a firm point. This is the origin of all the demiurgic dreams of scientists. Second, the power of a mathematical demonstration is greater than that of all other contrary experimental, political, or common sense evidence. This is the origin of the distinction between rhetoric or sophistics on the one hand and apodeictic reasoning on the other. Third, a well designed piece of machinery may reverse the balance of forces and change for ever the face of the Earth. This lesson will not be lost on long lineages of Princes and Engineers who will enlist non-humans on their sides to win over the multitude of humans. But, fourth, all the little gadgets that empirical sciences may produce are nothing compared to the pursuit of disinterested and unnecessary knowledge. As a go d Platonist, Plutarch shows all the scientists of the future how a little bit of abstract reasoning can be made more valuable, more real and more weighty than any earthly achievements. We could perhaps add a fifth reversal: a tale of an experiment that stages the demiurgic power of a scientist is stronger than any historical account showing the contrary. This is still felt by historians of science.

If you combine these four reversals of forces as intricately as the pulleys attached to the merchantman, you may at once do politics without doing any, build machines without building any, strike this world and its masses without ever leaving the world of ideas. Indeed, it is the very notion of a Platonic Idea that sums up all the ultimate reversals of forces to which Archimedes and Plutarch are subscribing: give me the Idea, the noumen, and I will move the world, the phenomenon; give me the essence of man, what Man *is*, and I will have more, know more, weigh more, than if all the empirical, real, unreal, shadowy, empirical men and women were given to me. What we admire in Platonism, this audacious reversal of reality and common sense, we should admire in a lever or, for that matter, in an atomic explosion.

The extraordinary power of Archimedes' demonstration is to render *commensurable* and *continuous* through a law of proportion the small and long arms of the lever. But, the extraordinary power of Plutarch's demonstration is to claim that although the science of the scientists and the politics of the King are made at once continuous and commensurable, they are nevertheless transformed, through a law of *dis*proportion, into an absolute *incommensurability* between the world of Reason and the world of Forces. For Plutarch a mathematical demonstration is not in continuity with a political demonstration, a show of strength, even though it reverses the outcome of any show of strength. Plutarch strikes twice, not just once: Archimedes sees that the only basic difference between the light and the heavy, the single and the many, the weak and the strong, is a law of proportion. This is enough to surprise the King and his retinue who believe in the reality of all those differences, but then Archimedes traces a yawning gap between the base world of the demonstration of forces and the high spiritual world of the demonstration of reason. This is enough to strike the King and his followers who started to believe in the continuity of politics and science! This is what makes this text so exemplary for a study of science and society.

When and where does Plutarch talk about politics? When Hiero comes into the story challenging Archimedes? When Hiero concludes the public experiment with a micro Manhattan project? Yes, but not just then. Politics comes into the picture before: when Plutarch explains why ideal mathematics are *superior* to vile empirical means (xiv, 3-6); politics is there when Archimedes' arm balances the weight of the fully laden trireme; and politics is still there when Plutarch, Plato and Archimedes contemptuously erase all traces of mechanical gadgets and earthly mechanics to return to a world that is infinitely stronger than this base one. If politics were limited to Kings and people, then there would be no need for public experiment, for compound pulleys, and for abstract demonstrations. But if politics consist in reversing the balance of forces, in reshuffling the definition of who is stronger than whom, nay, if politics consist, since Archimedes, *to compose, mix and confuse, to make both commensurable and incommensurable various balances of forces*, then we understand why Archimedes is on the beach this beautiful Sicilian afternoon, applauded by the crowd, baffling the King with *tes tecnes ten dunamin* [See Figure 1].

'He declared that, if there were another world, and he could go to it, he could move this one'. Well, he does have this other world, he lives in it, and it is from here that he moves the merchantman, the King, the Syracusans, Marcellus, and the whole lot. This is the world of science. It is the world of Ideas where Archimedes is firmly entrenched, his fulcrum at hand, experimenting on how to move our world. The Platonist realm so much vaunted and so much denounced, is not spiritual, is not detached from ours. Rather they are in continuity, they bear the same relation to each other as the little weight on one arm of a scale to the vastly greater weight that is balanced out on the other side. In our admiration (or our hate) for Platonist Ideas, we too often forget that it is first of all a fantastic machinery to reverse all relations of forces by tying science and politics in a huge 'minimax' relation. A vastly stronger force *is* a force. A force which is so different that it is not even engaged in the base politics and technics it can reverse at will, is a force all the more so.[2] The whole mystery of the field of Science,

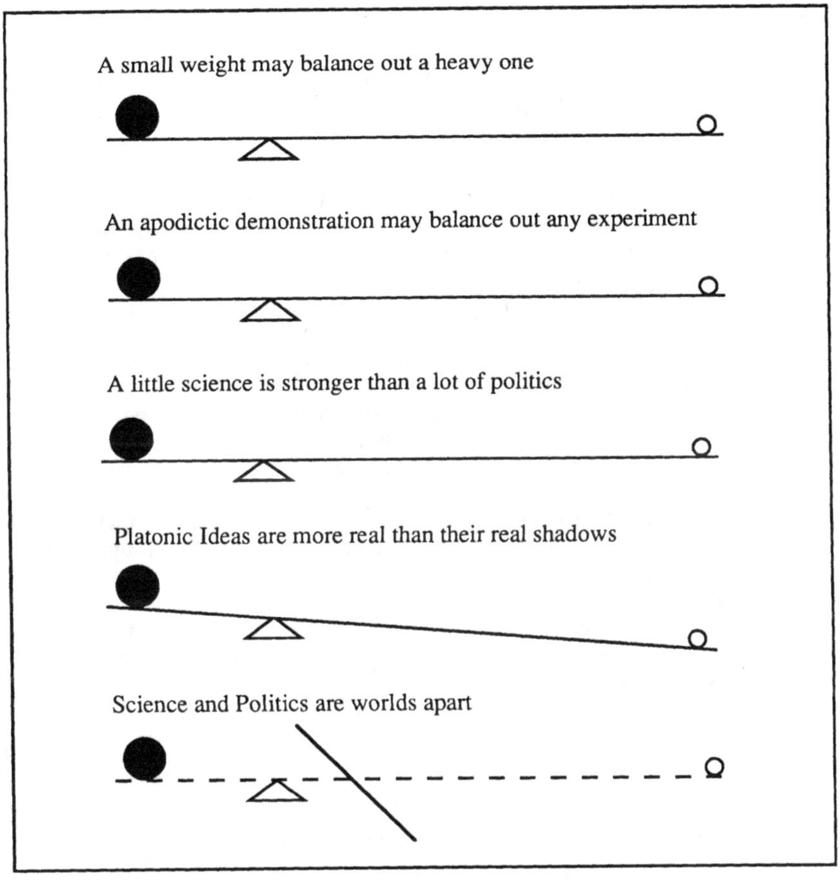

FIGURE 1

Technology and Society Studies hinges on this balance of forces made continuous and discontinuous, commensurable and incommensurable at the same time.

2. PERON ON THE TASMANIAN SHORE

Baudin, the French explorer, more fortunate than Archimedes, goes from worlds to worlds, and together with many Cooks and La Perouses, allow the European nations, through the tiny lever of geography, sailboats and expeditions, to raise quite a bit of this Earth. Peron, the official zoologist of the 1802 Baudin expedition, is a closet anthropologist mostly concerned with interviewing wary aborigines on the beach as a way of solving questions raised by Rousseau about the inherent goodness of the Noble Savage:

I then wanted to try to make use of the dynamometer I had brought with me. I thought that the form of the instrument and the way of using it would perhaps attract their attention and occupy them for a few moments. I was not mistaken. They admired the instrument, all wishing to touch it at the same time, and I had most difficulty in preventing them from breaking it. By making repeated trials with it in their presence, they seemed to understand how to use it, and already four had done so, and I was hoping that all the rest would follow their example, when the old man of whom I have spoken and who until then had remained silent, retaining always on his face a profound look of misgiving, distrust and wariness, uttered a few words which, without having the air of being precisely an order, produced however an effect such that no one of them dared to touch the instrument again. Undoubtedly it would be difficult to know precisely what this old man actually said to his compatriots, but from the attention he had given to everything I happened to do with the dynamometer, it was evident he had understood its use perfectly. Then, reflecting on the object we could have for such an operation, his reason and experience had been able to tell him: 'without doubt, these strangers only wish to test our strength in order to use the information they acquire against us in future'; and his compatriots, whom he made acquainted with his reflections, having understood that his suspicions had a real foundation, did as he wished, seeming even to keep up some resentment about what I was doing' (Plomley, 1983: 89)

The aborigines become still more restless when Peron, abandoning the suspect dynamometer and hoping to calm them, reverts to the classical barter system and starts exchanging spears for beads and bottles. The anthropological interview soon turns into a show of force that would not have favoured Peron had it been pursued (90-1):

The old man had filled the hearts of all with his suspicions and his fears: 'These treacherous strangers,' he must have said to them, 'want to know your strength only so as to come afterwards to oppress you by the strongest possible means.' Such talk could not fail to excite against us a general mistrust, and this mistrust seemed to become certainty from the moment when we sought to obtain their weapons *at great cost*. After that it seemed we were only seeking to deprive them of them, so as to strengthen the execution of our treacherous designs.

Even if Peron grant the old man a conception of a balance of power and strength relations which is unthinkable for the aborigines, the way he aligns the dynamometer and the exchange of goods -- micro-scale experiments -- inside the bigger scale experiment of European exploration is most revealing of *his* notion of power. Just as Baudin's ship tests the relative strength of the coastline, inscribing on a map the movement of their ship (and hardship) to decide what is worth knowing, having and colonizing, Peron tests the relative strength of the savages he meets to decide whether or not civilisation is a degradation or an improvement. Every navigator does the same by bartering goods, trinkets and trophies to test the relative value of everything on earth compared to everything else. Peron does not try to hide the link between his objective science, the market forces, and the objectives of the expedition. Not only does he tie them to one another but he *grants* the aborigines the *same* appraisal. They too are portrayed as sociologists of science seeing a power relation beneath the data and the exchange. To do so they are helped by the fact that the dynamometer itself -- wrists or loins against lever and needles -- and the barter system, spears for bottles -- are literally relations of strength.[3] Peron's assessment of the situation or the assessment he supposes is being made by the old man, are both quite extraordinary given that neither could predict in 1802 that this little experiment would end up by wiping out the Tasmanians altogether. Before colonization has even started, the story is terminated by Peron and attributed to the foreknowledge of the savage. 'They will oppress you by the strongest possible means'. This little island will indeed be moved by the fulcrum of civilization, but the aborigines will not follow the movement. They will be shaken out.

But, in the report that Peron writes in his *Voyage*, it is the Noble Savage thesis that is wiped out much more quickly than the real savages. In his chapter entitled 'Experiments on the savage people of New Holland', Peron (150), after comparing Tasmanians with French and English sailors, is forced to conclude that Native Man is weaker than Civilized Man: 'the inhabitants of Van Diemen's Land [Tasmania], the most savage of all, *the true children of nature*, are the most feeble' (Peron's italics). 'That memorable epoch has not so long passed when we saw celebrated men, trapped within a vivid imagination and embittered by the misfortunes inseparable from our social condition, denounce it because they failed to recognize the benefits, and reserve for the savages all sources of happiness and all virtuous principles' (146). Here goes Rousseau, forced outside the debate by the strength of Regnier's dynamometer and the computation of Peron's measurements. In addition to the spring and needle of the dynamometer, in addition to the cartographic expedition of Baudin, there is a third balance of forces, that between arguments. Rousseau speaks without data. Peron has the data behind him expressed in kilograms. He has been there. He has tried the dynamometer on the beach and recorded the readings, although not as many as he wished because of the savage's critical

sociology of science. Experimental anthropology should replace the weaker discourse of philosophers as surely as the settlers should displace the weaker aborigines.

This anecdote allows us to see the difference between being *experienced* and being an *experimenter*. Dozens of navigators had experienced the coast of Australia, mostly by dying on her reefs; dozens of sailors had experienced the savages by marrying them, killing them, selling them or being pierced by their spears. The skill, knowledge, and know-how thus acquired exists, but incorporated, embodied in the very flesh of the experienced person. Bits and pieces of Terra Australis started to be experimented upon, when experienced navigators recorded each encounter in a common written or visual language and shipped these records to a common place where they could be gathered and combined. But instead of being incorporated, their skill, knowledge and know-how is *excorporated*, inscribed and turned into papers and maps piling up inside the walls of some scientific or commercial or political institution. You have to have been *through* the trial and then *out* of it. Before and after this slight shift from body to inscription, sailors suffer hardship, wreck their boats, die of dysentery, and feel the exhilarating beauty of the Tasmanian coastline. Before and after they are tried by the coast and by the natives. But, now they have turned their bodies, their crew, their ship into the inked needle of an instrument of enormous proportions that scribbles the shape of Van Diemen's Land in London, Paris or Den Hagen. And while they are recording the shape of the land, even before shipping it to Europe, in the crowded cabin of their boat, surrounded by stuffed specimens of birds and dried plants, they are using this freshly made inscription to make points against their colleagues, Captain Cook, Marion Du Fresnes or Rousseau with whose work they have surrounded themselves. Peron, like all the scientific navigators of the late eighteenth and nineteenth centuries is travelling both *to* Tasmania and *within* the scientific literature. He cannot record a piece of data on his dynamometer without seeing the article he is going to write reversing the point made by another writer, back there in France, about the Theory of the Noble Savage. Each recorded experiment is fodder for thought, for the agonistic field back home.

This is why there is so little difference between *observation* and *experiment*, as often noted by Claude Bernard. An observation is an experiment where the body of the scientist is used as instrument, complete with its writing device, that is, a hand, a quill and a notebook. Peron, before testing the strength of the Tasmanians and the quality of their sociology of science, undergoes all sort of trials in the aborigines' hands. He barely avoids exposing his most intimate parts. After a sailor had performed a few sleights of hand, a savage pierces him with a needle to see if blood comes out. He allows them to smear him with charcoal. 'We were so novel to one another! The natives wanted to examine the calves of our legs and our chests, and so far as these were concerned we allowed them to do everything they wished, oft repeated cries expressing the surprise which the whiteness of our skin seemed to arouse in them' (84). When undergoing these trials, Peron *notes, records and writes*; his colleague, Petit, *draws and paints*. The savage tests the whites' testes, but they do not write down the result, let alone make every effort to send it away. He, Peron, availing himself of his new friends' nakedness, writes a long chapter on the absence of erection among the Tasmanians! Both parties gain knowledge through experience. But, while the savage gains experience of the white man, the scientists, offering their bodies to science, turn themselves into an object of experiment. It does not matter if he has an instrument -- the dynamometer, his own body -- prodded by the savages whose reactions he registers accurately, his eyes -- trained, biased and disciplined, or the huge laboratory-like paraphernalia of clocks, projective geometry and cartography that Baudin carries on board. Only three things matter: that there is a trial, that there is an inscription, that there is a point in the literature to be made or unmade. With these three things taken together Peron is confident that his time and the money of the doomed Republic has not been wasted, that Rousseau's Noble Savage argument can be put to rest, and that his data will back him up in case of controversy.[4]

This iron tie between trials, inscriptions and fields is more important than all minor distinctions between experiments that are man-made and observations that are supposed to confront phenomena on which man has no control -- coastlines, stars or dancing cranes, or with field expeditions in the open and expeditions of probes and meters inside the protected walls of a laboratory. One could say that Tasmania is 'observed' by Baudin and that the strength of savages is 'experimented upon' by Peron, because the former is merely sighted without being manipulated, whereas the latter is a highly elaborated construction depending on dynamics, instruments, protocol and (inaccurate) statistics. This, however, would be a misleading distinction since the whole of Baudin's expedition is the moving and registering part of a larger instrument whose 'screen' or window is set up in Paris, and whose data are feeding various discussions on the size of the strait between New South Wales and Tasmania or on the possible colonization of South Australia. These gentlemen back in Paris experiment on what the Earth looks like, they bet and theorize, waiting for the results of the expedition, as surely as if they were microbiologists in their laboratories waiting for a culture to grow. It only happens that their laboratory has the size of the Earth, and that the probe they send away is manned by men and relies a lot on disciplined eyesight, but it could be an unmanned satellite, or a core drilling.

Visually, the inscribed results brought back from an expedition, from an observation, or from an experiment look alike and this is what counts in feeding the discussion.[5] They 'refer to' which means, etymologically, that they 'bring back' things. Archimedes was building a continuous relationship between the small and large weights, through the calculated proportion of the lever. Each instrument builder, each expedition, each experimenter sets up a balance very similar in shape to Archimedes' enterprise: the small paper-like window of the inscriptions balances out what it refers to -- the vast world 'out there' -- through the maintenance of certain proportions (Latour: 1986) [See Figure 2]. Anthropologists,

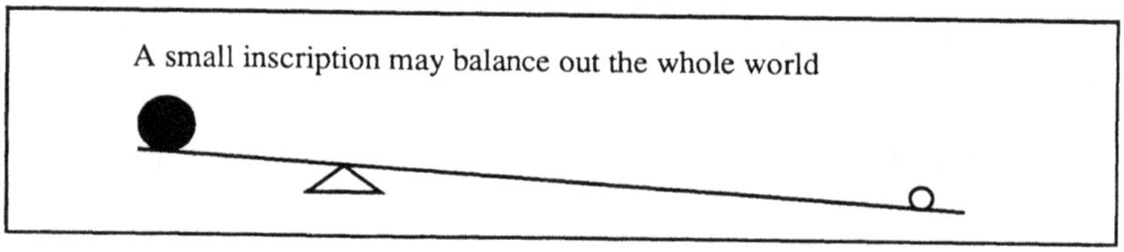

FIGURE 2

geographers and zoologists in Paris can now indulge the two favourite gestures of realist philosophers, *i.e.* thumping the table ('this is a hard fact!') and pointing at details with their index finger (the other meaning of reference). But, what they point to and thump on are the *inscriptions* brought back through whichever means the ingenuity of scientists has invented. In the aula of the Institute, in the cabinet of the Societe des Observateurs de l'Homme, there will be discussions about these thin and delicate pieces of paper, and it will be as if they were talking on, and talking of the Tasmania shore, not thumping on their table but on the beach, not pointing at the statistics but at *this* old aborigine himself. The inscriptions are only the thin, almost immaterial, sharp end of a huge instrument. And as miraculously as on the Syracusan beach, the gentlemen shuffling around this side of the lever will be able to reverse the 'vastly superior' forces of Tasmania and the Tasmanians.

However, for this new balance of power to be enforced, it is necessary for the overall apparatus to possess another paramount property: the traces that are brought back must not be *made-up*. Baudin has to prove that he has been there. Peron has to prove that he has been there. No matter if they deal with man-made or naturally occurring phenomena, no matter if they deal with phenomena inside or outside the laboratory, what counts is that the human discussant has behind him data that do not depend on man. How can this unreliable and querulous political animal introduce into the discussion things that are non political, that are reliable, and that put discussions to an end?

3. PASTEUR ON THE VERGE OF FERMENTATION

Thus, there was a great interest in deciding whether racemic acid would undergo the same fermentation as the right tartrate acid, in other words, whether the yeast would transform the left tartrate acid as readily and in the same way as the right tartrate acid. The ammonium racemate was fermented in the same manner as I indicated above for the right tartrate. The fermentation manifested itself as easily, with the same characteristics and deposit of yeast. But in following the development of this phenomenon with a polarizer, it was clear that things are altogether different. After a few days of fermentation, the liquid that was at first inactive, now possesses a discernible rotary power on the left, and his power increases progressively during the fermentation until it reaches a maximum. At which point fermentation is then stopped. There is not a trace remaining of right acid in the liquor which, once evaporated and mixed to its volume of alcohol, provides at once an abundant crystallization of left ammoniac tartrate.

No doubt this offers an excellent means to prepare left tartrate acid. But the whole interest of this fact seems to me to depend on the physiological role of fermentation which appears, in my experiments, as a phenomenon pertaining to the vital order. We see here that the character of molecular dissymmetry so peculiar to organic matters intervenes to modify the affinity. There is no doubt that it is the type of dissymmetry characteristic of the molecular arrangement of the left acid tartrate that is the only and exclusive cause of the non-fermentation of this acid in the very conditions where the symmetric acid is destroyed.[6]

In his laboratory in Strasbourg, Pasteur is *designing* an *actor*, an actor as new as the exact shape of the Tasmanian coast designed, drawn together, by Baudin's expedition. How does he do this? By defining trials for the actor to show its mettle -- a metaphor coming from another trial of strength. Why is an actor defined through trials? Because, there is no other way to define an actor other than through its action and there is no other way to define an action but by asking what other actors are modified, transformed, perturbated, or created by the character that is made the focus of attention (Latour: 1988). This is a pragmatist tenet, which is extended to the thing itself soon to be called a 'microbe', to the story told by Pasteur to the Academicians and to the reactions of Pasteur's interlocutors to his story.

Pasteur is engaged at once in *three* trials of strength that should be first *distinguished* and then *aligned* onto one another. First, in the story told by Pasteur there exist characters whose (semiotic) competence is defined by the performances they undergo. The 'hero', placed in front of a tricky choice between two different but deceptively intermixed outcomes, is at first puzzled and then, without hesitation, sorts out the two alternative branches, and triumphs, applauded by the elated audience. One can think of the story of Joan of Arc recognizing the King of France at Chinon in the middle of a crowd although He had disguised Himself and given His attire to another man but one can just as well think of the yeast able to select out, molecule after molecule, the right from the left handed version of the racemic acid that was designed to fool it. In both cases, the hero has undergone the difficult trial with success and defined part of its competence: 'she is really inspired'; 'it is really a living organism'. As far as stories go, 'Joan of Arc' and 'Clever Yeast' are comparable actors, although one is anthropomorphic and the other zoomorphic [See Figure 3].

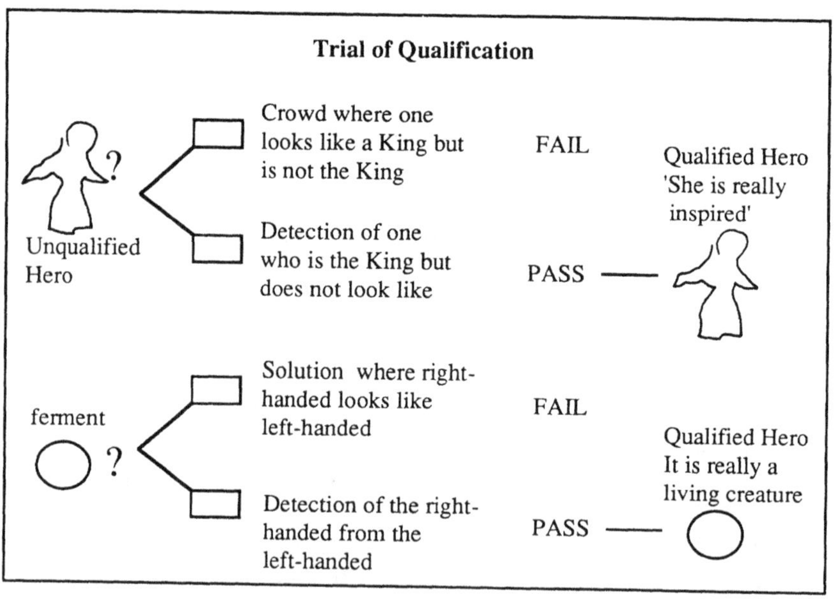

FIGURE 3

Second, Pasteur in his laboratory is busy staging a new artificial world in which to try out this new actor. He does not know what the essence of yeast is. In his laboratory he is a good pragmatist: essence is existence and existence is action. How the hell does this actant behave in this medium (right-handed tartrate) and in that medium (racemic acid)? What does it produce (fermentation)? What does it break down? What it is up to? Most of an experimenter's ingenuity goes into devious plots and careful staging designed to make an actant participate in new and unexpected events that will then actively define it. He, Pasteur, had also been tried out by racemic acid a few years previously. An objection had been raised against his earlier work in crystallography. Pasteur triumphed by sorting out the left crystal from the right with tweezers under a microscope, thus showing that the racemic acid was a combination of the two forms in equal proportion. This was by now a non-problematic feature of his laboratory, know-how and equipment. Could the ferment be as clever as Pasteur's own hands and eyes? Could it have an activity similar to his? Why not? Pasteur invents a setting in which the yeast will be asked to do what Pasteur did: sort out the left-handed from the right-handed. The result of the experiment is to grant the would-be actor a sorting competence similar to that of Pasteur, only much quicker. Who is this little guy able to do in a moment what Pasteur has been doing at great cost?

The first trial is a story -- it pertains to language and is similar to any trial in fairy tales or mythologies. The second is a situation -- it pertains to non-verbal, non-linguistic components (glassware, yeasts, Pasteur, laboratory assistants). Or does it? The third trial is designed to answer this very question. Pasteur undergoes this new trial when he tells his story of the Clever-Little-Yeast-Who-Can-Tell-Its-Right-From-Its-Left. At the meeting on 29 March 1858, Pasteur is now trying to convince the Academicians that this story is not a story, but that it has occurred independently of his wishes and imaginative ability. To be sure, the laboratory setting is artificial and man-made but the competence of the yeast is *its* competence, in no way dependent on Pasteur's cleverness in inventing a trial that allows it to reveal itself. What happens if Pasteur wins this new (third) trial? A new competence will now be added to *his* definition. Pasteur is the one who has shown that yeast was a living organism; just as the second trial added a new competence to yeast: it can tell left from right. What happens if Pasteur fails? Well, the second trial has been wasted. Pasteur entertained his peers with the tale of Clever Yeast that was amusing to be sure, but which comprised his own expectations and earlier prowess [See Figure 4].

An experiment is, however, none of these three trials. It is the *movement* of the three *taken together when it succeeds or separated when it fails*. No experiment can be studied by being just in the laboratory, in

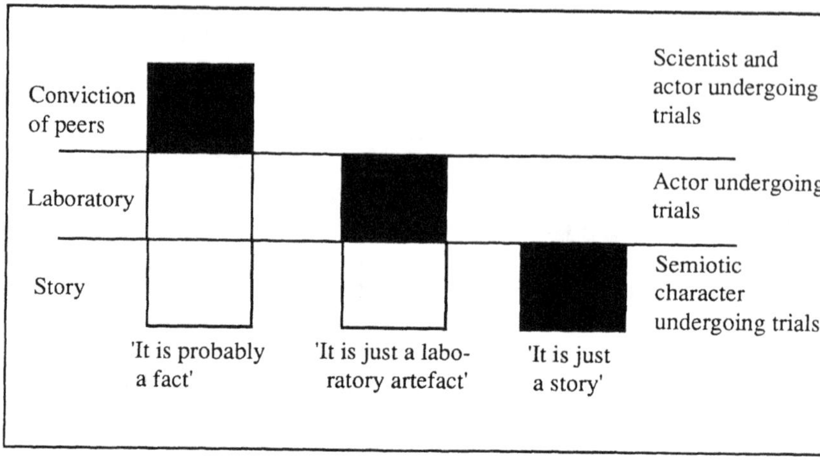

FIGURE 4

the literature or in the agonistic field. It is a story to be sure and studiable as such but a story *tied* to a situation where new actants are undergoing terrible trials plotted by an ingenious stage manager; and then, the stage manager is undergoing terrible trials at the hands of his colleagues who try out what sort of *ties* there are between the first story and the second situation. An experiment is a text about a non-textual situation tested by others to decide whether or not it is a text. If the final trial is successful then it *was not just a text*, there is indeed a real situation *behind* it, and both the actor and its authors are endowed with a common new competence: Pasteur has proven that the yeast is a living thing; the yeast is able to discriminate right from left. *They mutually exchange and enhance their properties*: Pasteur helps the microbe to show its mettle; the microbe, Pasteur to win one of his many medals. If the final trial be lost, then it *was just a text*, there is nothing behind to support it, and neither actor nor stage manager has won any *additional* competence. Their properties cancel each other out and colleagues can conclude that Pasteur has simply prompted the microbe to say what he wished it to. If Pasteur win we will find two (partially) new actors on the bottom line: a new yeast and a new Pasteur; if he lose, only one, he, the Pasteur of old, will be inscribed in the book together with a few shapeless and wasted chemicals. This triple definition of an experiment explains its two most prominent features as recognized by many analysts: first, it is both similar to literature yet wholly different; second, it is both artificial and real.

4. 'UN FAIT EST FAIT'

Scientific literature involves a very peculiar type of exegesis since it comprises a text that comments on another text -- as all exegesis does. But the latter is generated by way of situations and settings that are given the properties of inscribing, blotting, scratching, staining and writing. The passage between the three-dimensional setting and the narrative is obtained by the process of inscription. But, in addition to this first peculiarity, the text thus obtained is meant not only to be read. It is meant to fuel a polemic that tests the safety of this very passage from narrative to setting as provided by inscriptions. Moreover, once the polemic is closed, the result is no longer a text, but instruments, products, beliefs, know-how, equations, or other laboratories.[7] Pasteur's colleagues start making left acid tartrate. They do not just talk about it, they do not just write about it. From now on they work and live in a world where yeast is a living organism telling its right from its left, exactly as Pasteur learned to live in a world where racemic acid was made of the superimposed left and right crystals. Experiments do words with things and things with words through instruments, inscriptions and controversies. This is why scientists who obsessively inscribe, write and dispute, may say, with good grounds, that literature, inscriptions and controversies (the fodder of social studies of science) are of no relevance. This double-talk is no proof of scientists' false consciousness: depending on the outcome of the three trials defined above, texts, traces and polemics may indeed *lose* their relevance.[8]

The second main character of experiment is their artificiality that has been noticed by most analysts and most recently by Pinch (1986) Hacking (1983) and Knorr (1981). For Hacking, reference is interference or, as Bachelard sums up, 'un fait est fait' ['a fact is an artefact' or 'a fact is fabricated']. This is the case first because it is selected out of millions of other possible set-ups; second, because it is constructed in between instruments and is invisible as well as unthinkable without them ('phenomeno-technique'); third, because it is always underdetermined and has to be overdetermined by theoretical expectations. To whatever brand of realism a philosopher wishes to cling, the fabrication of phenomena in laboratories has to be accounted for. It is not even possible to overcome the problem by saying that artificial situations are there only to *reveal* or to dramatize the competences of phenomena which 'out there' would retain those properties quite independently from the artificial set-up. This 'out-thereness' has been shown to be always dependent on the *extension* of a laboratory situation which supposes, at least in the beginning, the risky replication in another setting (Collins, 1985). There is no way to *demonstrate* that yeast selects the left acid tartrate without borrowing from a laboratory the polarimeter, the yeast culture, the know-how of crystallographers and microbiologists. Of course one can *believe* that yeast retains this property outside the laboratory but this is a belief, a respectable one, but not a demonstration. To be sure, the set-up may be simplified, routinized, black-boxed, so that it appears to spread effortlessly to many places but, if looked at carefully, the ties that hook up the black box to laboratories are always as visible as the cable that hooks a computer to the mains. No spread of laboratory artefacts will ever prove that the same (what could it mean?) phenomena (what could they be?) would be observed (by whom?) outside the laboratory (where that could be?). 'When a rationalist insists that behind the facts there is the *ground* of facts, the *possibility* of the facts, the tougher empiricists accuse him of taking the mere name and nature of a fact and clapping it behind the fact as a duplicate entity to make it possible' (James, 1907: 263)

All this is well known. It has been centuries now since a correspondence theory of truth has been articulated by anyone with the slightest acquaintance with the building of laboratories. And still, lo and behold, whatever we think or argue, the left acid tartrate is *not* invented by Pasteur, but by the yeast. At least, this is the very problem for his colleagues, for himself and for the little bug down in the crystal solution. It is essential to all of them that whatever the ingenuity of the experiment, whatever the perverse artificiality of the set-up, whatever the underdetermination or the weight of theoretical expectations, Pasteur manages to extract himself so as to become an 'experitus', that is, someone transformed by the manifestation of something not itself contrived by the former Pasteur. No matter how artificial the setting, something new, independent of the setting, has to get out, or else the whole enterprise is wasted. It is because of this 'dialectic' between fact and artefact, as Bachelard puts it, that although no philosopher defends a correspondence theory of truth it is absolutely impossible to be convinced by a constructivist argument for more than three minutes. Well, say an hour, to be fair. Most philosophy of science since Hume and Kant consists in dealing with, evading, hedging, coming back, recanting, solving, refuting, packing, unpacking this impossible antinomy: facts are experimentally made up and never escape from their man-made settings; it is essential that they are not made up and that something emerges which is *not* man-made. Bears in cages pace back and forth on their narrow turf with less obstinance and less distress than philosophers and sociologists of science going from fact to artefact and back.

This obstinance and this distress are due to the definition of an experiment as a zero-sum game. If an experiment be a zero-sum game, if every output must be matched by an input, then nothing escapes from a laboratory that has not been put into it. Whatever the philosopher's list of the inputs in a setting, it always features the *same* elements before and after: the same Pasteur, the same yeast, the same colleagues, or the same theory. Unfortunately, since it is at once fabricated and not fabricated, there is always more in the experiment than was put in it. So, *explaining* the outcome of the experiment by using a list of stable factors and actors will always show a *deficit*. It is this very deficit that will then be accounted for differently by the various realist, constructivist, idealist, rationalist or dialectic persuasions. Each will *make up* the deficit by introducing their favourite stocks: Nature 'out there', macro- or micro-social factors, the transcendental Ego, theories, or the nice melting pot of dialecticians. There seems to be an endless supply of fat bank accounts upon which one can draw in order to complete the list and 'explain' away the originality of an experiment. In this kind of solution, the novelty is not accounted for by modifying the list of initial actors, but by adding one paramount factor that balances the account. In this way every input is equilibrated by an output. Nothing new has happened. Nature (or society, or theory, or *x*) has simply been *revealed* at the occasion of an experiment.

But, nothing proves that an experiment is a zero-sum game. On the contrary, every difficulty above suggests that *an experiment is an event*. No event can be accounted for by listing the elements that enter the situation *before* Pasteur launched his experiment, *before* the yeast started to eat up the right handed tartrate, *before* the meeting of the Academy. If such a list were made, the actors would *not* be endowed with the competence that they will *acquire* in the event: Pasteur is a promising crystallographer but he has not shown to anyone's satisfaction that the ferments were living creatures; the yeast may accompany the fermentation, as Liebig claimed, but is not yet endowed with the property of selecting out left- from right-handed crystals; as for Academicians they do not depend on a living yeast in their own laboratories but prefer to remain on the solid ground of chemistry. The list of inputs does not have to be completed by drawing upon any stock resource, since the one drawn upon *before* the experimental event is not the same as the one drawn upon *after*. This is precisely why an experiment is an event and not a discovery, an uncovering, an imposition, a synthetic *a priori* judgement, the actualization of a potentiality, and so on. This is also why the list drawn after the experiment shows no deficit whatsoever: a (partially) new Pasteur, a (partially) new yeast and a (partially) new Academy, are all congratulating each other. In the words of Isabelle Stengers, an experiment should not be explained by 'the principle of sufficient reason'.[9] The reasons brought to bear on the explanation are insufficient, not because one factor has been forgotten, or because the list has not been carefully drawn, but because actors *gain* their definition through the very trials of the experiment. It is always admitted that science grows through experiment: the point is that Pasteur is also modified and grows through this experiment, as does the Academy, and, yes, as does the yeast. They *all* leave their meeting in a different state than which they went into. In other words, there is a history of science, not only of scientists and there is a history of things, not only of science.[10]

This proposition itself has to be put to the test. There is no better example than the Tasmanian coast, chosen above as a paradoxical case of experiment. If it can be shown that such an obvious counterexample is a collective and continuous historical event then *a fortiori* it will be the case for facts/artefacts that are clearly laboratory-made. In what sense is Baudin's voyage an event that should not be accounted for simply by a list of stable actors? There is no difficulty in realizing how deeply the voyage is an event for Baudin since he dies in the process. It is also an event, although a lesser one than which was expected, for the commissioners of his expedition back in Paris; their maps of the Southern Hemisphere will never be exactly the same afterwards. But in what sense is Tasmania *herself* an actor modified in the trial?[11] That Tasmanians will never be the same afterward is clear from the elder's perceptive appraisal of the dynamometer; they are all doomed. That the landscape will be deeply and lastingly modified, is also clear, terribly clear to anyone now driving through the Verdun-like landscape of deforested land. But what about the coast, the land mass, the shape of the land, in what sense can these be said to be partially constructed and partially realized during the experimental event of Baudin's mission?

First of all, a coastline is also an actor in the definition given above; *i.e.*, it does things. For instance, it has the nasty custom of wrecking ships, of hiding itself in mist and rains. It is also unpredictable, as much as a cloud shape. From the shape of this cape no one on earth could deduce that there will be a bay, an island or a river estuary behind except, that is, the *second* navigator who reads the map drawn by the first and 'deduces' from the characteristic recorded shape of 'Presqu'Ile du Nord' that, for sure, the 'Riviere du Nord' lies behind. To establish the first map and make the life of the second navigator easier, certain coded details of the coast have to be gently experimented on by a ship coming close -- not too close; staying away -- not too far; moving gently -- not too slowly, not too fast; and taking bearings that have to be carefully recorded, summed and projected. The coastline and the landmass are laboratory events; they are unpredictable at first and then slowly and painstakingly stabilized. To hold this cape down there in the mist, to make it out as a characteristic shape, to introduce it in the sequence of movement that leads you to the 'Riviere du Nord', you need to enter it, like any other object, as a written answer to a succession of trials (what longitude? what latitude? what height? which profile when approached from the east?).

Is it possible to jump *out* of the experiment being made by the ship to the Tasmania out there? No, because to tell *what* is out there or even whether this is Tasmania or not, you have to rely on just this laboratory work. Baudin happens to this cape as much as the cape happens to Baudin (Law: 1986). It is given a name, it is recorded, extracted, characterized, stabilized, sequenced. No one could say that it was there before Baudin, without pointing at or thumping on another, earlier map, and since it is Baudin who made the first map, this would be difficult! The only way to argue that it was there before and independently of Baudin and of the Dutch and English cartographers would be to point at a *shapeless* mass of land that would *not* take into account any of the details experimented on by those expeditions. Thus, going back in the past, undoing what each navigator has done, erasing the result of each map in turn, the obstinate realist will be left pointing at a shape that has become so shapeless that it will not be called Tasmania any more but Terra Australis Incognita, of which nothing can be said. In the end, he or she will be left pointing an index finger at 'It There'. Either you fill the reference 'It There' with something specific and you have to get nearer the collective work of cartographers and administrators and colonizers, or you stay away from this work and are left with a reference to nothing in particular. You can't have your reference and eat it too.[12]

This impossibility of escaping from collective work does not mean that Baudin invented Tasmania out of his head. Or, at least, this nagging doubt is slowly being rejected by the many actors who raise for themselves this very philosophical question: is Baudin reliable? How real is his map? Less and less can people argue that Tasmania was invented by Baudin, since in Paris, in London and at Botany Bay, different maps are compared, superimposed, and create an *internal referent* that settles the discussion: Baudin has not made it up. Other navigators are adding to his definition. Tasmania's coastline starts to stabilize there in the cartographers' cabinets and in the colonial offices by calibrating the laboratories (Jardine: 1986). Whereas it had been texts and reports, actions and hardship, heated discussions, it is now a black box. The result of the experiment is now sunk deeper in other types of practice, instrument, deduction, and cliche. From now on, they all learn to live in a world to which is added a Tasmania of *that* shape [See Figure 5].

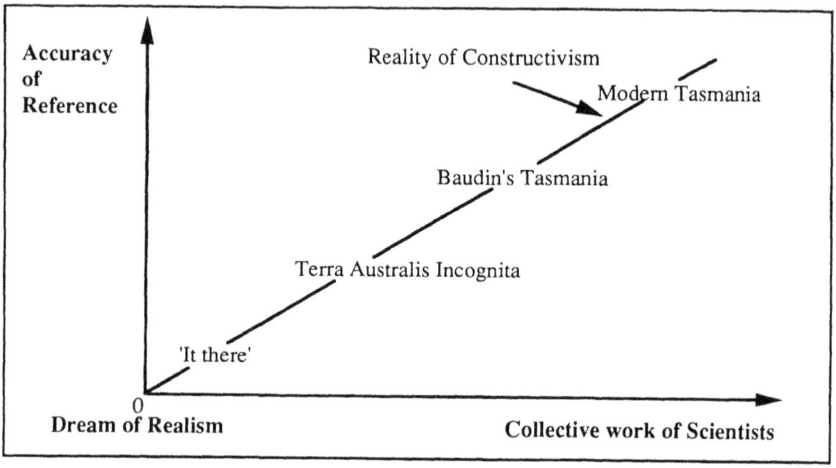

FIGURE 5

But, what about Tasmania herself? She is no longer herself, that is the point. Like the sleeping Gulliver she is now tied solidly by a great number of tiny but tight links fastened by those Lilliputians. She can't move and disappear. More and more frequently fleets are coming to her shores. For her coastline to become stable and reliable, for the map of Tasmania to look like Tasmania, capes are equipped with signals and light houses, towns are marked by names and signs, so that now more and more people sailing 'there' may establish a 'correspondence' between the map and the coast. The laboratory situation of the cartographer is now brought back to the place to which it 'refers'. Words like 'correspondence' and 'reference' are not mysterious and forbidden terms. They do not designate the useless *extremes* of Reality and Representation, but the collective work *in the middle* which builds reversible translations between inscriptions on the scaled-down map and scale one Tasmania (see Brannigan, 1981 on Columbus' work). They now designate the *extension* of laboratory conditions to stabilize the actors' definition and to render it *durable*. It is this passage from words to things and from things to words that is so difficult to follow for those who oppose Nature and Signs or Reference and Interference. The two extremes (Representation and Reality) are not the most important parts of this network of actions and events and no one has ever to account for the face-to-face meeting of these two equally shadowy figures. The whole reality-representation is spread in the middle and along this network no one in particular is more important than any other. They all compose this on-going event: Tasmania. And, this event is not yet achieved. You, too, happen to Tasmania as much as she happens to you.

If I go to Tasmania in order to solve this question, I realize indeed that the Tasmania where I am standing now is *not* on the map of Tasmania. This eucalyptus half hidden in the mist, this flying wild goose, this creek, this fern, this destruction are not inscribed on it. When I am out there it is not on the map and when I read the map the place where I stand is not there. To be sure, there are connections in certain crucial points between names on my map and names that I come across: 'Devonport' is on the map and I also see a sign 'Devonport 1 km'; the distances in kilometres between Devonport and Hobart are the same on these signs as they are on the map. I can superimpose these two readings and they will more or less fit with one another. But this correspondence does not count as one between Representation and Reality since they are both coming from the same source -- cartography, administration, surveying -- and since the signals and posts have been fixed in the ground to turn a few landmarks into annexes and complements of the map. If you destroy the signals and ask me to test the relation between my map and this extent of land, I will start to be lost. I may ask someone 'Is this Devonport?' But, now let us take out, one after one, each inhabitant whose collective experience grounds the map and its landmarks solidly into Tasmania. What happens to me when I land 'there', with no one to ask 'where' is Devonport or whether indeed this is Tasmania or Australia or King Island? I have to become a geographer, a cartographer, and explorer. So I have to do the work of mapping all over again. My present weight is little compared to that of the island? But, this is because I am a tourist coming after two hundred years of visits, actions on and exactions from Tasmania. Although a visit to her shore transforms me more than it transforms her I still happen to her, a little bit; I am a Lilliputian part of her construction.[13] She is the reference provided you add to her definition the work of all of those who ground her, and the instruments and embodiment of all of those who are still acting to make her exist and survive. This continuous and collective historicity of *all* actors when they enter into a new trial solves the antinomy of fact and artefact -- they are made up; they are not made up.

The question raised by realism (does this acid or this coast correspond to something out there?) may now be separated in its two *components*: a true passion and a wrong solution [See Figure 6]. There

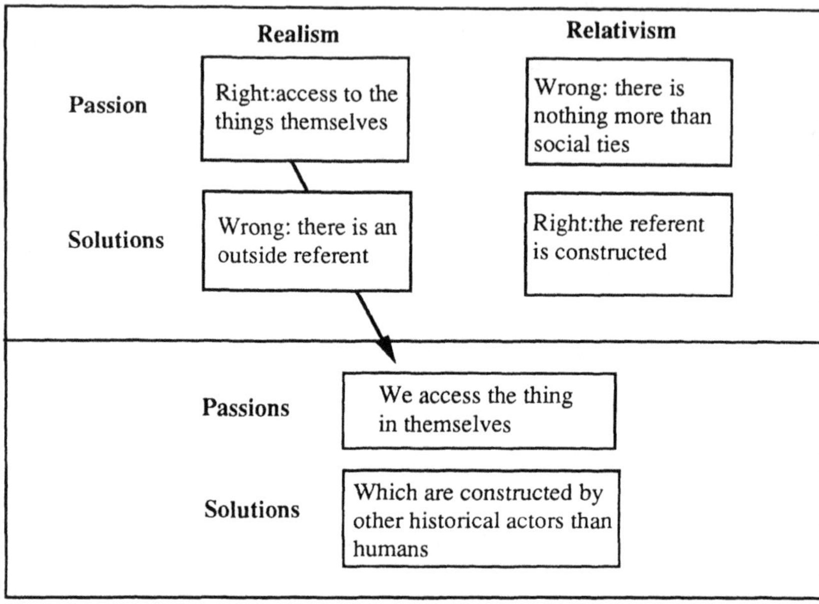

FIGURE 6

is among realists a passion for accessing the things themselves, for building the world with more than social ties, with more than human minds and this passion should be retained. The wrong solution to fulfil this passion is to claim a fanciful repetitive projection of what the sciences do into what the world is, and that solution should be ruthlessly discarded (Woolgar: 1988).[14] The same distinction may be brought to bear upon the various brands of constructivism: it is a practical and pragmatic solution to a wrong passion; laboratory settings and experiments are indeed historical and local achievements; but the passion to reduce them to a *stable* list of theoretical, cognitive, or social human factors is unnecessary. More precisely, the polemic passion of constructivists is triggered, understandably, by the absurd solution of realists, and this solution in turn is seen, quite understandably, as the only way to escape the constructivists' solution; hence the Tom and Jerry type of chase in which they all so much delight. But, there is a way out which consists in choosing the pragmatic solution to express the right sort of passion: *a little bit of constructivism takes you far away from realism; a complete constructivism brings you back to it*, even though the population of actors having a common history will never resemble the old Out-Thereness. Thus constructivism is not to be confused with construction by society. In building our world we (man? mind? society? ideas? labour? episteme?) are far from being the only actors. You can have your cake and eat it too, or, to remain amidst gastronomic metaphors, there *is* such a thing as a free lunch.

5. PRAGMATOGONY

> The most striking feature of the consultations that we are going to study is the following: when the *kindani* [diviner] 'works' for the preparation of public ceremonies he puts himself in the hands of the king who is mainly responsible to the society for the fortune or misfortune bestowed by the invisible powers that govern human destiny.... This is why the consultations are made on a grand scale. They may last one, two or three weeks.... They are held under the direct control of the King of Lere who is constantly informed of the partial results and of the problems encountered. If, for instance, sacrifices are necessary to modify a statement which is especially worrying, the King provides the animal (chicken, goat or sheep) whose blood will be spread where the draws are made, after which the meat is eaten by the diviners.... When the *kindani* is completed, the King comes to the consultation arena to have the results read and commented. He then offers gifts to the *pa-kindani*: mil, tobacco, beer, silver.
>
> Having the shape of a large imperfect circle of around thirty meters in radius, the *tehale* [sacred enclosure] is a depression surrounded by *tebakame* (balanites aegyptiaca). Inside this circle are also planted a few of these trees at the foot of which are seated the diviners to work with their stones or more exactly with the special stones that never leave this *tehale*. Big river pebbles which are amassed around will be used to report the final results on the outside circles. At the end of the *kindani*, these results will be recapitulated and discussed in front of the King and his retinue who will then have, if there are two rows of stones inscribed on the soil, almost four hundred meters of 'sooth-saying text' to inspect (Adler & Zempeni, 1972: 153).

Here is another King, not Hiero of Syracuse but the King of Lere in Chad. Here are other specialists, the *pa-kindani*, who *work* in an *insulated* enclosure with *specialized* tools (stones and pebbles which are thrown at random), wasting laboratory animals in the process (not rabbits and rats, but chickens and goats). Here is an *inscription device* of great proportions where each result is carefully recapitulated into rows of other stones, the disposition of which are endowed with technical meaning. Here is another public *experiment*.

It will be remembered how Plutarch cleansed Archimedes of all attachments to this base world of mechanics after he had impressed the public and defended Syracuse. Here too there is a rite of erasure. In the final day of the *kindani*, the King's horsemen race to the consultation arena and dismantle the sacred writings with the hooves of their horses, while the diviners wash themselves and cast a spell on the horses' hooves, passing to them the responsibility for the spells that were inscribed in the stones: 'we are not the cause of the misfortunes here inscribed' they say. Whereafter the horsemen rush to the river and cleanse their mounts of any accusation. Nothing has happened. The arena is neutral again. This being the end of the public experiment, the new year ceremonies may begin (185-6).

Syracuse, no doubt, was full of such experiments at the time of Archimedes' trial; so was Tasmania before the aborigines had to submit to Peron's silly dynamometer; so might have been the city of Strasbourg while Pasteur was busy teaching tricks to his yeast. All of them, scientific or prescientific, have the common feature that they are man-made and yet still introduce into our commerce elements that do not depend on our will and human intention. It was the mathematical world of demonstrations that Archimedes was bringing to bear upon the usual political affairs; as for Baudin or Pasteur, their colleagues became convinced that they had not made up their whole stories; that they were indeed bringing into our world a new bit of Tasmanian coast and a new actor, the living yeast; and that these were here to stay with us as part of our world. The *pa-kindani* also insist that they are *not the ones doing* the inscriptions. It is at random that they throw their stones. They do not bias the throw. They do not cheat. It is at the occasion of this non-human, non-social, non-political draw that the powers that govern our destiny indicate what they want us to do. The diviners are as surprised by the results, as worried by them, as the public might be. As any good scientist would do, they ask the King for new sacrificial animals in order to *replicate* the surprising statement and to check whether or not they can withstand another test. The data are so independent of their will that they wash and decontaminate themselves, in the end taking no responsibility for the inscriptions, but instead throwing them to the horses.

Could their honesty be doubted? Yes, but Plutarch's, Peron's or Pasteur's experimental or story-telling abilities could also be contested. Maybe the *pa-kindani* skew the throws of the stones to manipulate the

King or the clans? Surely, but Archimedes manipulates Hiero as effectively as the trireme; Peron's critique of the Noble Savage is highly distorted; as for Pasteur, he will soon move the whole of France with his microbes. So what is the difference, if any? To pin-point this difference, it might not be necessary to doubt the Moudang's honesty, or to invent some form of 'savage mind', or belief in witchcraft, or to redefine any deep Great Divide.[15] Even if all our characters, scientific and pre-scientific, are assumed to be honest experimenters in contact with non-human actors which are not their invention, a difference remains:

> The primary function of the *kindani* is a complete inscription answering a complete demand of the society.... During the *kindani*, after having checked whether or not the supernatural powers, the village and its land, are in favourable dispositions, the question is to spot and control people.... The *kindani* is simply an enumeration of all the characters who, for one reason or another, are part of the representation of the ceremony... The control of people in the preparation of the collective ceremony aims at verifying their ritual capacity: are they in the required state of purity and integrity so that, because of them, the community is not put in any danger now that it has entered a ritual period? (183-4)

The *kindani* is presented -- in as much as we can believe Adler and Zempeni's report more than Peron's statistics! -- as a huge *checklist* of all the relevant aspects of the society, before another year begins. Each notable, each village, each clan, each of the king's wives, granary and organs, each of the government executives, is inspected one after the other with a throw of stones and required to do something if the answer is dubious or threatening -- amend or sacrifice. The divination, the stones, the rites, the independence of the diviners, are employed to underline, mark, test and repair the ties of the social fabric, very much as a fisherman would do with each mesh of his net before setting sail. Is the social structure disguised by the rite? Are the diviners really using their knowledge of their society under the pretence of non-human throws of chance? Is their independence a mockery? No, because the social structure is in part made by this check list, repair work, and close annual inspection, just as the fisherman's net, after years of inspection and repair, ends up being a new net. There is no difference between building and repairing. The non-human stones, the independent diviners, and the transcendental power that talks through chance, are necessary components of the society. No society of humans exists without the non-humans to hold it together.

So what is the difference between Archimedes on the beach and the Moudang in the middle of their stones? What is the difference between what we call an experiment, a scientific one, and what we, academic

Westerners, no longer call experiment? After all, in both cases there exists a genuine appeal to non-humans and a true independence of the experimenters overwhelmed by realities they do not control, even though each have a clever hand in fabricating them? The answer is to be found in the *movement* of the non-humans and in the *direction* their spokesmen impose on them as a way of reinforcing *or modifying* the fabric of social links.

A 'thing', etymologically, is an assembly, usually a judiciary assembly in the course of which an accusation is made: the root is the same in Latin, 'causa' and French, 'chose' (Serres, 1987). Thus, the first origin of a thing is a collective assembled to accuse and probably to sacrifice. Serres calls 'pragmatogony' the slow movement that leads through a series of substitutions from a purely social and collective definition of the 'thing' to a definition that requires more and more 'objects' to hold it together. You start with a collective; you end with a collective *plus* a nature, *plus* technique. You start with a judiciary cause; you end with a scientific cause. You start with an accusation; you end with a causation.

At each step of the (necessarily mythical) pragmatogony, non-humans are mobilized, enrolled, lifted out of their environment, to be brought to bear on social links, but their movement, their aim, their 'angular velocity', so to speak, is not the same. In the *kindani*, for instance, the non-humans are used by the diviners to trace, or retrace the already established but constantly weakening social ties: family lines, clans, land tenures, purification rites, power structure. These ties would not be visible without the passage of these 'tracers'. The tracers play the same role as the ball in a game of rugby. The ball seems to be what the game is about, since everyone tries to reach it, but as soon as one holds it, it is got rid of and passed to someone else. In the end, it is the trajectories of the successive *passes* that define, *in actu*, the collective. Similarly, the stones thrown at random are cleverly directed by the diviners to each of the relevant points of the society that is completely mapped out in three weeks and inscribed in rows of stones. Of course, the diviners are not doing a directly social analysis, but anyway such a 'directly social analysis' is the Western anthropologist's understanding of the whole rite. For the Moudang, the randomness of the stones, the mediation of the diviners, the invocation to supernatural powers are necessary elements of the analysis just as the ball being sought and shifted is a necessary component of the game of rugby. Could teams 'directly' run against one another and bump into each other and fight without a ball? Could diviners 'simply' tell who is powerful and who is not, who is sick and who is dangerous? No. The non-humans (balls, rules, stones, gods) are necessary for the collective to exist. But, since these non-humans are neither objects nor subjects, natural or social, Serres calls them *'quasi-objects'*. Are they here simply to trace, reveal, mediate, reinforce the whole collective? No, they are there to turn the scattered assembly into a whole, to repair the decaying collective. They are there to collect the assembly, to perform the society. This is why their *trajectory* is essential. It is crucial that they be both non-humans and completely collective, able to trace all the details of the collective without forgetting any member or articulation. If they stop, they lose meaning. If they be human, too human, they lose meaning. Are the diviners of the *kindani* cheating? Certainly yes, they have to, but it does not matter. In three weeks the whole society has been mapped out, inspected, checked up, repaired, reconciled, ready for another go.[16] Who did this? No human did it. The hooves of the horses have erased everything and have then been cleansed of the spell.

Archimedes' lever, Peron's statistics, Baudin's coastline, Pasteur's yeast, circulate very differently through the collective. They too are non-human. They too are man-made. They too are brought to bear on the collective. But what do they do, once brought in the collective? *Locally* they *create a unanimity* and, once this unanimity is ensured, they *reverse the relations of forces* elsewhere in the society. Their mode of circulation does not turn them into quasi-objects. We are now much further into the (mythical) pragmatogony. They have become *objects*, things, black boxes no one has made and which are going to change everyone's world. The quasi-objects were living in our social world. We are going to live in a natural or technical world of objects. There was one world, *made* one by the circulation of quasi-objects. There will be many worlds spread apart by the circulation of objects.

Much has been made of the ability of experiments to put otiose discussions to an end. 'Settle matters of fact first, and useless chats, polemics and verbosity will be stopped', thus speaks the scientific wisdom. But, this interruption of debates is carefully circumscribed as it has been so beautifully shown by Shapin and Schaffer (1985) in the case of Boyle's experiments. To be sure, discussions about the validity of Archimedes' demonstration are no longer possible among the King's retinue; debates about the strength of Primitive Man die out in the meetings of the Paris Societe des Observateurs de l'Homme; opposition to the idea of living yeasts peters out under the Academy cupola. But, what about the rest of the social fabric? What about the complement of this set? It is transformed, disrupted, destroyed, reshaped, by the quasi-miraculous intervention of non-humans accompanied by their spokesmen. The age-old relations of force are reversed by Archimedes' pulleys; the Tasmanians are wiped out; convicts are moved to another hemisphere; breweries are transformed by microbiology. The paradox of experimentation is that the locally achieved consensus is paid for elsewhere by a refuelling of dissensus on a new and larger scale. The grey boxes of the *kindani* could not reach further than the very society it mapped out. The black boxes of the experimenters allow action at a distance. It is because the controversies are locally settled by matters-of-fact, that the rest of society may be moved out of its usual ways. Indeed, the very relation between the few settings where matters-of-fact are settled and the rest of society reproduces the very shape of the lever.[17] A little weight moves a great one: 'Give me a laboratory and I will raise the world'.

Do not try to raise the world with quasi-objects; you will not have the leverage and you *will not have another social world from which to modify the first*. You can do whatever you want with quasi-objects, but there is one thing you cannot do and that is to reverse the relations of forces in any grand way. You can alter, modify, repair the social ties, but not make a gaping hole in them and replace them by non-humans. Convinced collectives of experimenters, however, will do just that. Quasi-objects have to be able to circulate everywhere to trace social ties. It is enough for objects to stop inside the laboratory walls. In order to circulate everywhere quasi-objects have to be continuously open to compromises, interpretations, negotiations. Objects, when they are pushed out of the laboratory settings, have to be uncontrovertible black boxes which no one can renegotiate. It does not matter too much if there be a doubt on the origin of quasi-objects; are they really coming from the gods, the diviners, the King? A large uncertainty is not only possible, but is necessary to allow the tracing of the social fabric. But it matters enormously, obsessively, uniquely, whether laboratory objects are man-made or not. The whole point -- the Archimedean point! -- is missing if it can be proved that the scientists have only other humans (interest, politics, affectivity, psychology) behind them rather than the non-humans they claim to reveal.

The peculiar anthropological trait of scientists is that they have to insist on the absolute distinction between non-humans and humans, and that, nevertheless, this locally achieved absolute distinction is immediately transformed into a modification (slight or deep) of our world. Like Archimedes, they establish a continuity, a commensurability, a law of proportion between reason and force, and then sever all thinkable links between force and reason. Truth is not an obsession of the *pa-kindani*; the circulation of quasi-objects through the society is, and so too is the turning of the society into *one single* society. But it can be safely concluded that truth is the obsession of the experimenter, as is the circulation of *objects* through one society that it turns into several.

Both types of societies resist the ethnographer's account of their similarities and differences with the same zeal. Other societies are furious when the rationalist Western ethnographer tells them that they need to distinguish between knowledge and society, between their culturally biased *view* of the world and what the world *really* is. We, in turn, are furious when the ethnographer of science, equipped with the tools and methods of the exotic fields, insists that we should make no distinction between our knowledge and our society, between what the world really is and the highly cultural ways in which we shape it. The first ethnographer imposes a *dualism* that threatens the society he studies; the second ethnographer imposes a *monism* that threatens the society in which he lives. These two impositions and these two flares of anger and passion are probably unnecessary. The clever anthropologist (who has first moved to study other cultures, then has come back to study ours, and who then ponders why so much passion is generated by these two types of explanations) realizes that the two might be equally misleading. He now sits in the middle of the Great Divide and learns that what defines one society cannot be used to define others: instead we have to see *our* dualism between science and society and *their* monism as the two major features of the two types of societies. The ethnographer of science who sits over the Great Divide is not confronted with prescientific and scientific worlds, but with societies which would fall apart if a distinction were made between their knowledge and their social fabric, between reason and force, and others, that is our society, which would fall apart if this distinction were *not* made.

ACKNOWLEDGEMENTS

I thank the History and Philosophy of Science Department of the University of Melbourne for the two month leave that made this paper possible. I thank Isabelle Stengers for her passionate understanding of realism. My friend Steve Woolgar who kindly corrected the English wrote 'I am not the cause of the misfortunes here inscribed'.

NOTES

1. This experiment is still the most popular and the most often staged in 'hands on' science museums. It offers little kids the nice occasion of reversing the balance of forces by dragging effortlessly their father, thus learning the easy way in what the social studies of science consists and why the accusation so often made against 'constructivists' and 'relativists' of relishing 'mere relations of strength' is so misleading. See Authier (1989).
2. As usual, the critiques of the 'constructivists' attribute this obsession for relations of forces to *us* when we are simply trying to give the philosophicalframework (Latour, 1988: Part II) that render understandable the scientists' obsession for establishing and then denying relations of forces. Killing the messenger is the strategy of those people who strangely enough accuse us of immorality!
3. 'Description et usage du Dynamometre' par le C.[en] Regnier. On Regnier and Peron see the detailed studies by Miranda Hughes (1988a; 1988b; this volume)
4. At least if the statistical challenge is not too strong since Peron manages to capture only 18 readings for the whole Southern Hemisphere.
5. They meet and combine with other paper forms which are visually, optically, and geometrically compatible (coming from thought experiments, from theoretical calculations, or from numeric simulations) in such a way that to do the work in the centre the origin of the various written sources becomes irrelevant except in case of doubt or controversy (see Latour, 1987: Ch. VI).
6. *Comptes rendus de l'Academie des Sciences*, seance du 29 mars 1858, XLVI: 615-8).
7. For all this movement see Latour (1987) and Callon, Law & Rip (1986).
8. This is enough to see the main difference with fiction. In the 'high' literature, a text is not about another one present in it under the form of inscriptions (internal/external referents); it cannot be summarized and never loses its relevance *qua* test (you cannot replace Proust's *Recherche* by an abstract without losing the *Recherche* but summarizing a long article is what it is made for); it is not normally inserted in a controversy among people who are the author's peers so as to test the ties between the narrative and the referent to which they subscribe. This huge difference in the regime of *enunciation* does not mean that scientific texts cannot be studied with the normal tools of text semiotics. As far as the *enunciate* is concerned, they are fully comparable. See Bastide (in preparation).
9. The principle of sufficient reason defined by Leibniz states that there is nothing in the cause that is not in the effect and *vice versa*. The two lists balance so exactly that the phenomenon under scrutiny may be made *reversible*: the effects may give the cause back. This is both the principle of determinism and the dream of scientific reason: give me the conditions, all the conditions, I will give you the effect, all the effect, and *vice versa*. But Prigogine and Stengers (1988) argue that this principle does not work better for physics, cosmology or chemistry than for the 'narrative sciences'. Irreversible events are not defined by their conditions of possibility.
10. The limit of pragmatism is to be concentrated on man (individual at that). But if essence is existence and existence is action, this pragmatism is to be extended to the things in themselves now endowed with a history. James was ready to 'add to reality'. He transforms the metaphor of the Book of Nature, from a book one reads to a book one *writes* ('on the pragmatist side we have only one edition of the universe, unfinished, growing in all sorts of places, especially in the places where thinking beings are at work. On the rationalist side we have a universe in many editions, one real one, the infinite folio, or *edition de luxe*, eternally complete; and then the various finite editions, full of

false readings, distorted and mutilated each in its own ways' (1907: 259). But, he was prepared to do it as you add shape to a shapeless and plastic matter, not as you meet other non-human actors who have also their history. This shift away from human overcomes the other limit of pragmatists. They are unable to explain the *enduring* quality of a settled dispute. They have no way slowly to withdraw existence out of essence. This withdrawal occurs by shifting the task of maintaining the consensus to non-humans and moving from interactions, talks and controversial practices to a world *in which* we live.

11. It would be all too easy to talk of plate tectonics so as to add activity to the shape and position of Tasmania but I would have to rely on another discipline: geophysics. I am trying here the hard case of 'purely descriptive' cartography.

12. This is why realists are never able to *deliver* the Reality their feeling for reality is seeking; paradoxically, this Reality will be provided by Constructivists who like to describe themselves as having no special craving for Reality (see below). It is because we like reality so much that we have all been engaged in the detailed description of the nitty-gritty of science in the making.

13. So much for the classic distinction between what is socially new and what is personally new, between the discovery of things no other human has found and what is learned by students and tourists. The learning of an age-old discovery is part and parcel of this very discovery even if the weight of the new discoverer is comparatively small.

14. The belief in a correspondence theory is easily explained, however, when one considers the *superimposition* of traces and inscriptions on which scientists rely to build a proof. It is perfectly accurate to say that we have no other way of proving than by subscription, that is, by underwriting one layer of documents by another one. This process, however, remains inside the laboratory. It builds the *internal* referent, not the external one. It is the mistake of the correspondence theory to take a genuine process inside experimental practice for the external cause of its solidity.

15. On this question of the Great Divide see Horton (1977; 1982). In his first article he distinguishes between open and closed predicaments in order to account for the difference between Us and They. Then, in his second article he rephrases his argument using primary theories (which are universal) and secondary theories (which are elaborated in our societies by rival schools and disciplines). The major point is that They try to maintain at all costs one 'single overarching frame' whereas We don't.

16. This is the limitation of Horton's excellent analysis of his Little Divide: he provides no mechanism for the 'one single overarching framework' to be *made* one, *made* overarching and *made* a frame. To do so one has to introduce the trajectory of quasi-objects, a mapping mechanism of some sort in order to perform and maintain the unity of the frame. Goody (1977) offers a much more tractable solution since he focuses on intellectual technologies.

17. Mary Hesse insists on the difference between metaphoric and metonymic relations to express the same distinction as the one between quasi-objects and objects. To take the part for the whole is, she says, the scientists' obsession. The part (scale model, laboratory, theory, maps, models, experimental fields) *is* the whole. In other terms, if instead of quasi-objects, immutable mobiles are made to circulate a different society is traced out and performed.

REFERENCES

Adler, A. & Zempeni, A. (1972). *Le baton de l'aveugle: Divination, maladie et pouvoir chez les Moudang du Tchad*. Paris: Hermann.

Authier, M. (1989). Archimede ou le canon du savant. In *Elements d'histoire des sciences*, ed. M. Serres. Paris: Bordas.
Bastide, Francoise. (in preparation) *La semiotique des textes scientifiques.*
Brannigan, A. (1981). *The social basis of scientific discovery.* Cambridge: Cambridge University Press.
Callon, M., Law, J. & Rip, A., eds. (1986). *Mapping the dynamics of science and technology.* London: Macmillan.
Collins, H. (1985). *Changing order: Replication and induction in scientific practice.* Beverly Hills: Sage.
Goody, J. (1977). *The domestication of the savage mind.* Cambridge: Cambridge University Press.
Hacking, I. (1983). *Representing and intervening.* Cambridge: Cambridge University Press.
Horton, R. (1970). African thought and Western science. In *Rationality*, ed. B.R. Wilson, pp. 131-71. Oxford: Blackwell's.
Horton, R. (1982). Tradition and modernity revisited. In *Rationality and relativism, ed. R. Hollis & C. Lukes.* Oxford: Blackwell's.
Hughes, M. (1988a) Tall tales or true stories? Baudin, Peron, and the Tasmanians, 1802. In *Nature in its greatest extent*, ed. R. MacLeod & P. Rehbock, pp. 47-72. Honolulu: University of Hawaii Press.
Hughes, M. (1988b). Philosophical travellers at the ends of the earth: Baudin, Peron and the Tasmanians. In *Australian science in the making*, ed. R.W. Home, pp. 34-63. Melbourne: Cambridge University Press.
James, W. (1907). *Pragmatism, a new name for some old ways of thinking.* London: Longman, Green, & Co.
Jardine, N. (1986) *The fortunes of inquiry.* Oxford: Clarendon Press.
Knorr, K. (1981). *The manufacture of knowledge.* London: Pergamon Press.
Latour, B., ed. (1985). *Les 'vues' de l'esprit.* Culture technique, **14**.
Latour, B. (1986). Visualisation and cognition: Thinking with eyes and hands. In *Knowledge and society: Studies in the sociology of culture*, ed. H. Kuclick. London: Jai Press.
Latour, B. (1987). *Science in action.* Cambridge: Harvard University Press.
Latour, B. (1988). *The pasteurization of France.* Cambridge: Harvard University Press.
Law, J. (1986). On the method of long distance control. In *Power, action and belief: A new sociology of knowledge?*, ed. J. Law. Sociological Monographs 32, London: Routledge & Kegan Paul.
Pinch, T. (1986). *Confronting nature: The sociology of solar neutrino detection.* Dordrecht: Reidel.
Plombey, N.J.B. (1983). *The Baudin expedition and the Tasmanian aborigines, 1802.* Hobart: Blubber Head Press.
Plutarch. (1961). *Lives.* Trans. Bernadotte Perrin. Cambridge: Harvard University Press.
Prigogine, I. & Stengers, I. (1988). *Entre le temps et l'eternite.* Paris: Fayard.
Shapin, S. & Schaffer, S. (1985). *Leviathan and the air-pump.* Princeton: Princeton University Press.
Serres, M. (1987). *Statues.* Paris: Editions François Bourin.
Woolgar, S. (1988). *Science the very idea.* London: Tavistock Publications.

Ecole des Mines, Paris

MIRANDA HUGHES

THE DYNAMOMETER AND THE DIEMENESE

I am the first to have opposed actual experiments and numerous facts to an opinion which is too prevalent, perhaps too dangerous and certainly too exclusive, that the physical degeneration of man follows the improvement of civilization.

With these words Francois Auguste Peron (1807) set his experimental results against the conventional beliefs of the eighteenth century. His 'actual experiments' involved the use of a dynamometer to measure loin and hand strength. The dynamometer became the lever to overturn the cherished Enlightenment doctrine of the Noble Savage. Rejecting the long-held assumption of a natural relationship between savagery and strength, Peron claims instead that the state of savagery is inherently a state of feebleness.

Labillardiere and Coulomb's (1804) report on Peron's 'Experiences sur la force physique des peuples sauvages de la Terre de Diemen, de la Nouvelle Hollande et des habitans de Timor' recommended publication in the Institut de France's *Memoires des Savans Etrangers*. This endorsement from the most prestigious scientific body in Europe indicates that Peron's experiments in anthropology well fitted the accepted scientific methodology and that no distinction was then perceived between experiments in physics and chemistry and those in the human sciences. Peron, through his application of the practices of experimental natural philosophy (see Schuster & Watchirs, this volume), sided with that discipline's concept of objectification through the use of data against the traditional approaches of Enlightenment 'anthropology' in which *philosophes* rarely ventured beyond the armchairs in a salon.[1]

1. THE PORTABLE DYNAMOMETER

The dynamometer measures the amount of muscular force exerted by a human or animal. The portable dynamometer was developed by the physicist Edmund Regnier in 1798 in response to a request from Buffon who suggested the need for a device that could be used in a wide variety of situations and would be less cumbersome than the available English

H. E. Le Grand (ed.), Experimental Inquiries, 81–98.

model. Regnier's model was not only compact but, unlike previous ones, the measurement of the muscular strength of both hands and loins could be done on the one instrument. In the same year of 1798, Coulomb (1800) presented to the Institut a memoir on the physics of the capacity of labourers for work. The appearance of (at least) two papers on the possibility of a 'science of work' indicates a nascent awareness of the role that physics and mathematics could play in creating measurements of human endeavours.

Measurement of human strength with the dynamometer continued to be in vogue during the first twenty years of the nineteenth century. Its popularity faded in the following decades, but appeared again in a modified form during the 1890s. The disappearance of dynamometer experiments for seventy years likely reflects the carving up of methodologies amongst the disciplines and the relegation of 'appropriate' experiments to different disciplines. A physical and apparently 'objective' measurement such as a dynamometer reading could be considered too strong, too accurate, for the human sciences, already downgraded to a class inferior to that of the physical sciences.

Regnier's 'Descriptions et usage du Dynamometre' appeared in 1798. The dynamometer is both described and illustrated in the article [See Figure 1]. Weighing about a kilogramme, the instrument measures

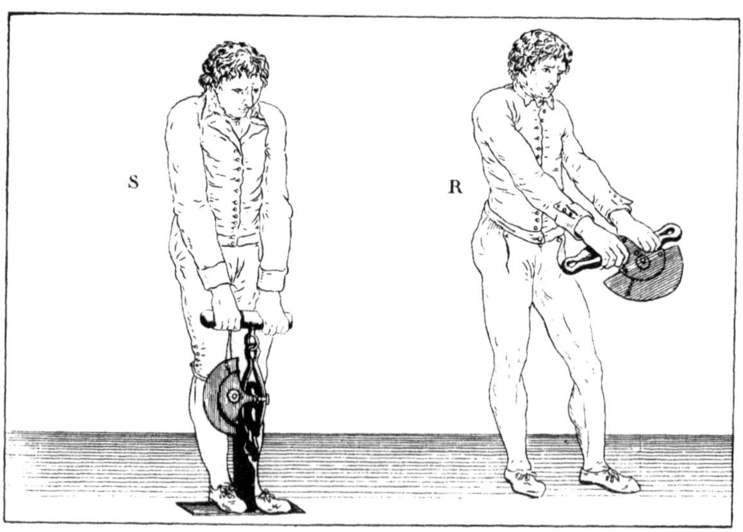

FIGURE 1: Regnier's Dynamometer (Regnier, 1798)

strength through the compression of a metal rod which causes a needle to register a reading on the dial. The design permitted a standardization of measurements performed using any of the many available instruments. That potential variation in measurement between dynamometers was mentioned indicates an awareness of the problems of transference of results, of the validity of ascribed objectivity of measurements. Regnier's observations and measurements came to constitute a standard scale for potential users. For a 'man of average strength' the mean reading for the hands is 50 kilogrammes, whilst for the loins and back, it is 13 myriagrammes. Although there are some 'very strong women', the average female reading is equivalent to a youth of fifteen or sixteen years, that is about two-thirds of an adult male (Regnier, 1798: 162, 166). Regnier also advocated the use of the dynamometer to judge the comparative strengths of animals in markets and sales. By attaching the dynamometer to a post, it was possible to measure the strength of horses and other beasts of burden (161).

Regnier also put forward several protocols and cautions to guide prospective users of the dynamometer. He notes that the optimal measurement of strength for the hands is obtained when the instrument is held at 45^0 to the body, and for the loins, when the body is held upright. However, he argues that the loin reading can give a false sense of strength because, like an architectural column, we are capable of supporting a considerable weight, that is, the weight-bearing capacity of a vertical post is more than one at an angle. Aware of the danger of conflating all classes of measurements, and to show that 'one ought not, however, always judge the strength of men by that of their wrists', Regnier provides examples of disparities between dynamometer readings and known work capacities of individuals. He also claims that measurements are only comparative, rather than absolute, influenced as they are, 'by our exercise, our manner of living and our customs' (168).

2. A VOYAGE TO SOUTHERN WATERS

In 1800 the dynamometer was listed among the 'instruments for navigation, astronomy and physics etc.' to be used in an official French expedition to the Southern Hemisphere.[2] That it was to be 'government issue' for the expedition suggests that it was by then seen as a standard piece of scientific equipment, an appropriate instrument for the task for which it was developed. On 19 October 1800 the *Geographe* and the *Naturaliste*, under the command of Nicolas Baudin, sailed from France and a slow passage was made to Île de France (Mauritius), Timor and the coast of western Australia. On board were twenty-four scientists, 'their knowledge and their baggage, for these gentlemen never move without pomp and magnificence' (Baudin, 1974: 442). Amongst the equipment was Regnier's dynamometer, and the savant who would make the most use of it, Francois Peron.

Peron, the experimenter and chronicler of the voyage, is the principal player in this saga. He had trained at the Museum d'Histoire Naturelle and the Ecole de Medecine. Amongst his teachers were Cuvier, Lamarck, Lacepede, Jussieu and Pinel. Peron was admitted to the voyage on the strength of a paper he submitted to the Ecole de Medecine, the arguments of which were based primarily on the medical model of his education. The main point was that knowledge of local diseases and their treatments would be beneficial to Europe. Following the doctrine of Hygiene, Peron (1799) argues that the two main areas of research ought to be an examination of the precise influence of the climate on the organic and moral constitution of a country's inhabitants, and an intense detailing of illnesses and native cures, recorded 'independently of all theories, of all hypotheses'. The natural history of man (as distinct from Rousseau's history of natural man) should also be investigated, and some attempt ought to be made to unravel the physical circumstances that contribute to their 'nearly unalterable state of health and nearly miraculous longevity'. The expedition's savants had already been chosen. Peron's rhetoric was, however, successful: he was added to the voyage, although in the class of 'zoologist' and not the requested *anthropologiste*.[3]

The twenty-four savants of Baudin's expedition (of whom only twelve reached the shores of New Holland in 1802) were to investigate all varieties of phenomena that would be encountered; *e.g.*, hydrography, geography, zoology, botany and meteorology. Baudin's voyage was the first to have had formal anthropological instructions. To guide anthropological investigations, the Institut commissioned four memoirs: the papers by Halle and Sicard are lost; those by Cuvier and the Ideologue lawyer and linguist, Joseph-Marie Degerando survive. Cuvier's paper (1800) signals the physical features that are to be observed and how to represent them artistically. However, it would be preferable to obtain physical examples of the different races, and Cuvier suggests returning with entire native families, at the best alive, but if this is not possible, at the least, preserved.

Degerando's treatise of some seventy pages, *Considerations sur les diverses methodes a suivre dans l'observation des peuples sauvages* (1800) suggests aspects of alien cultures to observe. For example, he writes apropos of their physical constitution and abilities (78):

> We shall be given more positive information about the physical strength of the individual savage.* You will discover what burdens he is capable of lifting, carrying or dragging; what are his most successful muscular movements; how quickly he can run; how far he can travel without rest; how good he is at swimming; what physical exercises he ordinarily undertakes; you will observe how he climbs tress, crosses ditches, climbs rocks and so on.

Rather than categorizing all savages as essentially interchangeable beings, Degerando advocates the individuality of each savage. In the note added to 'individual savage', he writes (78n):

> When we say the *individual savage*, the *savage*, the *savage people*, it will be understood that it is not our intention to speak of the savage in general, not to suggest that savage peoples are all of a single common type: this would be absurd. It is simply a short way of mentioning the *individual savage* or the *savage people* under the observation of explorers at the time.

Although not explicitly mentioned, one could safely assume that Degerando would see the use of the dynamometer as entirely appropriate for answering some of his questions - for example, physical strength - though it would not be the only way of determining even these capacities. By itself, the dynamometer could not provide sufficient information to cover all the required areas of anthropological research. A much broader understanding of foreign societies was required.[4]

3. 'THE SIGHT OF THE DYNAMOMETER'

Peron wrote little concerning the actual experimental procedures employed in connection with the dynamometer. Of measurements on the English and the French, no accounts appear to exist, despite the stay of five months at Port Jackson being covered by a description of forty pages in the *Voyage*. The largest number of readings were recorded on Timor but we do not learn how they were taken. Details of those on the aborigines of New Holland and Van Diemen's Land are mentioned incidentally but the focus of the narrative is elsewhere. In the chapter specifically concerning the experiments, little is said of the method of obtaining measurements: the text concerns the results and their interpretations.

The story of how the French experimenter and the Diemenese subjects reacted to the dynamometer trials is recounted in Peron's unpublished manuscript 'Maria Island: Anthropological Investigations' (Peron, 1800). Peron was not, of course, performing experiments on an inanimate, passive sample. Indeed, he did not *just* perform an experiment. The experiment, if we so describe the taking of these measurements, was one of many acts in the drama of encounter. Contact between the French and the Diemenese generated curiosity on both sides. In some ways, no clear distinction can be drawn between 'experimenter' and 'subject'. Both sides performed and waited for reactions. The Diemenese investigated the suspect physiques of their strange-coloured visitors whilst the French described what they perceived to be the myriad faults of aboriginal anatomy. Diemenese women covered the French faces with powdered charcoal; Frenchmen tried to embrace and kiss the native men.

But, there was always suspicion. Each side feared treachery from the other. The French, knowing that their own intentions were honourable, write of their dismay that the Diemenese did not trust them. Encounters often bordered on chaos when no common conventions could be maintained. Into one such atmosphere of uncertainty Peron introduces the dynamometer and attempts to gain readings. The Diemenese of Maria Island were already agitated by the approach of another French ship's boat.[5] In this tense atmosphere Peron decided to test the power of his instrument (1802: 89):

> I then wanted to make use of the dynamometer I had bought with me. I thought that the form of the instrument and the way of using it would perhaps attract their attention and occupy them for a few moments. I was not mistaken. They admired the instrument, all wishing to touch it at the same time, and I had much difficulty in preventing them from breaking it. By making repeated trials in their presence they seemed to understand how to use it, and already four had done so, and I was hoping that all the rest would follow their example, when the old man...retaining always on his face a profound look of misgiving, distrust and wariness, uttered a few words which, without having the air of being precisely an order, produced however an effect such that not one of them dared to touch the instrument again. Undoubtedly it would be difficult to know precisely what this old man actually said to his compatriots, but from the attention he had given to everything that I happened to do with the dynamometer, it was evident he had understood its use perfectly. Then reflecting on the object we could have for such an operation, his reason and experience had been able to tell him: 'without doubt, these strangers only wish to test our strength in order to use the information they acquire against us in the future'. Whatever may be the case, I have no doubt at all that they had formed this opinion of us and the experience will make me careful in the future as the use to which I will put this instrument. Very far from striving to make them understand its use, I will neglect nothing which will divert their suspicions while making allowance for their ignorance.

This encounter was not over. Peron tried 'in vain to gather further words' of the Diemenese dialect, and the French tried to buy the aboriginal weapons, spears and waddies, offering glass beads and buttons, but the Diemenese appeared to be loathe to part with their only means of defence. Peron postulates some very plausible reasons for this reaction, yet characterizes their response (1802: 91) as 'resentment...sudden and bloody', unable to concede that their behaviour may be rational. Peron and his compatriots feared an attack on retreat, but the natives disappeared into the bush. 'Since then we have not been able to see any more of them, and their fires have completely disappeared from all this part of the coast.'

4. ALL FOR ONE AND ONE FOR ALL

In 'Experiments on the physical strength of savage people of Van Diemen's Land and New Holland, and of the inhabitants of Timor' (Chapter XX of the *Voyage*), Peron writes (1809: 313-4) of the utility of Regnier's invention:

> Undoubtedly, although not perfect, and not able to furnish precise measurements of absolute strength, this apparatus at least gives a meaningful approximation to it. It produces comparable results and, for the numerous experiments I have made over the last three years, I am certain that in this matter it is much more exact than one could have imagined at first.... [I]f circumstances have not permitted me to develop this section of my investigations to the full extent to which it was capable, at least I always did my utmost to ensure that indispensable asset, exactness.

He produced a table (1809: 314) summarizing the mean measurements for each group [See Table 1]. This table, which gives no indication of the spread of results, was the only one presented in the English translation.

	Force des mains (kilogrammes)	Force des reins (myriagrammes)
Terre de Diemen	50.6	00.0
Nouvelle-Hollande	51.8	10.2
Timor	58.7	11.6
Francois	69.2	15.2
Anglois	71.4	16.3

TABLE 1: Results Generaux des toutes les Experiences dynamometriques

Accompanying this table are Peron's three main conclusions:

1. The inhabitants of Van Diemen's Land, the most savage people of all, *the true children of nature*, are the most feeble.
2. Those of New Holland, who are scarcely more civilized, are more feeble than the inhabitants of Timor.
3. These last, in their turn, are much more feeble, both for the hands and the loins, than Englishmen and Frenchmen.

Before presenting the more detailed versions of the results and conclusions that appeared in the French edition, it is important to note what Peron judged to be the soundness of his conclusions. He claims he had 'established by exact experiment that there is a degree of feebleness which is very remarkable' in both the aboriginal and Timorese groups. Peron prefaces his suggestions for the supposed reason of this feeble strength with an acknowledgement that a 'rigorous solution' to the cause of this will require 'more data than I have assembled here'. He also admits that, for the Diemenese, he was able to obtain only 'a few data and even those incomplete', but argues (1807: 148):

Nevertheless, all my observations having been made on those individuals of the soundest constitution, and the results with them having been quite definite, and above all, quite constant, we can, without fear of mistake, apply them to the generality of the individuals of this primitive race, the results indicating in all a deficiency of vigour which is truly extraordinary.

Peron here moves from a measurement of muscle tone to a conclusion about the overall health of the individual. However, the dynamometer only measures the power of a group of muscles and this bears only a sufficient, not a necessary, relation to the state of health.

The hierarchy of strength, according to Peron's three conclusions, is Diemenese, New Hollanders, Timorese, and, equal first, French and English. Peron uses as experimental subjects twelve Diemenese, seventeen aborigines from Port Jackson, fourteen Englishmen, including Lieutenant Governor-General Paterson, fifty-six Timorese and seventeen Frenchmen. The aborigines and the Diemenese were separated as 'different races of men' on the basis of differences in skin colour, hair texture, head shape and the proportion of the torso to the limbs. The Timorese were treated in the numerical tables as an homogeneous group but identified as three different races: Malays, Chinese and Timor's original inhabitants (Peron, 1807).

Peron includes 'comments' [see Table 3] about the physical constitution of his specimens. It becomes apparent from these that the descriptions are tied to the expectations and assumptions of the science of physiognomy. For example, one of the 'weaker' New-Hollanders,

Aca-ra-da-Pa-ra, is described as 'very feeble constitution; physiognomy hard and ferocious'. In contrast, some of the 'weaker' French were 'small and stout, but well constituted', or 'small, but well-shaped'.

Whilst the *average* results of Peron's tables may show a hierarchy of strength from Diemenese to European, an examination of the individual results reveals a larger spread [See Table 2]. If the physical constitution

LOIN MEASUREMENTS

Myriagrammes:	0	10-14*	15-19	20-24	25-29	30+
Anglais	3	-	1	6	3	1
Francois	-	-	5	8	4	-
Timorese	-	8	24	2	-	-
N. Hollande	-	9	8	-	-	-

HAND MEASUREMENTS

Kilogrammes:	30-39	40-49	50-59*	60-69	70-79	80+
Anglais	-	-	-	7	6	1
Francois	-	-	-	8	8	1
Timorese	-	3	15	14	2	-
N. Hollande	1	3	7	2	-	-
Diemenois	-	5	6	1	-	-

TABLE 2: Results of loin measurements (in myriagrammes) and hand measurements (in kilogrammes) by number of individuals in specified ranges. *The values for Regnier's 'average man' were 13 myriagrammes for loins and 50 kilogrammes for hands. Adapted from Peron, 1807.

of the individual is also taken into consideration, as Regnier advised, then the most appealing conclusion is that the apparently weak have weaker readings than the apparently strong, and this is consistently so for all races [See Table 3]. Indeed, Regnier (1798: 166) had alerted his readers to the likelihood of this observation: 'In general, men in whom the muscles are more pronounced are stronger than those in whom the limbs are fleshy like those of women'.

More puzzling are the actual numbers presented by Peron. The Timorese and aborigines fall entirely within the ranges Regnier designated as 'average Frenchmen', whilst the small sample of French and English that he tested were apparently capable of extraordinary strength. Part of their prowess was undermined, however, when Freycinet, who completed the narrative of the expedition after the death of Peron, discovered that Peron had apparently read the wrong dial

(Peron, 1807: 460-4). After adjusting the loin measurements, the set of

Names	Hand	Loins	Description
Nouvelle-Hollande:			
Paramara	60	12.4	Vigourous constitution, very long limbs, cruel look
Mon-guean	32	9.1	Also well-proportioned in the body; hair on all the body
AcaradaPara	44	13	Very weak constitution: physiognomy hard and ferocious
Van Diemen's Land:			
Baro-Ourou	53		One of the best built and most beautiful individuals of the nation
------	44.0		Legs and arms very feeble
------	60.0		Torso very robust; limbs feeble
Timor:			
Barimba	47.0	7.9	Big and fat; bloated figure; ignoble and crass.
Iatti	66	15.1	Big and fat; stupid physiognomy; ignorant and crass
Neas	55.0	11.7	Ancient King of Coupang; expression of an honest & delicate soul. Also well constituted
French:			
Michaux	82.0	18.5	Gaunt but robust body
Bailly	60	11.7	Short, large, fat, stocky; muscular system but little vigour
Moreau	72.0	13.0	Large, gaunt and meagre body
English:			
Harris	78.0	17.8	Short, very stocky, very strong
Jemisson	79.0	16.4	Large, big, strong and well constituted.

TABLE 3: Selected Examples from Peron's tables

readings fall closer to Regnier's estimate and to the results of an experiment on sailors at Le Havre which gave an average loin strength of 14.2 myriagrammes [See Table 4]. As well as this error, which affected all the measurements taken, there were two other factors influencing the range of values recorded among the races. The disparity in readings is, in part, due to the relative knowledge of the experimental subjects. The French and English know that the aim of the experiment is to appear strong and know how to achieve high readings. The Diemenese, on the other hand, in no sense could fully comprehend the overall meaning of the experiment, let alone such fine points of dynamometer methodology as holding the instrument at 45^0 to the body.

LOIN MEASUREMENTS

Myriagrammes:	0	5-9	10-14*	15-19	20-24	25-29	30+
Anglais	3	-	2	8	1	-	-
Francois	-	-	9	7	1	-	-
Timorese	-	7	26	1	-	-	-
N. Hollande	-	6	7	-	-	-	-

TABLE 4: Corrected Loin Measurements. *Sailors at Le Harve, 14.2 average. From Peron & Freycinet, 1816.

The design of the dynamometer could also lead to its incorrect use. After each pressing the needle remains fixed so that an accurate reading can be obtained. If any repeated squeezing of the apparatus be done, each previous reading is augmented, which could easily result in an overly large degree of strength being recorded. If the experimental subject tentatively squeeze the instrument, the needle will be inadvertently moved. Faced with the difficulty of trying to get the natives to test their strength at all, it would perhaps be expecting too much of Peron to check that the needle was at zero.

Peron writes of the validity of going from the few individual results to generalizations about the population of aborigines and Tasmanians in his general discussion of the dynamometer readings. In his 'Experience sur la force physique des Francois', he takes this argument much further. Its underlying theme is a justification of the small sample of readings taken, both for the French and the English, and for the aborigines. In fact, completely opposite reasons are given: few aborigines were measured because no individual differences are assumed to exist between them while too few Frenchmen were measured since they were considered as being such a disparate group than any results used to characterise them would be misleading. Peron argues this quite convincingly, and his rhetoric on the matter is worth quoting in full (1807: 414-6):

Among savages, or little civilized people, the dynamometrical observations offer a character of rigour and generality that one could never give them in the midst of our European society. Our numerous population, the diversity of our professions, of our exercises, of our food, of our habitual work etc. excludes all exact comparison; and for finding the mean term of the force of a powerful and civilized people, like the French, for example, one must make a very considerable number of observations, one must repeat them on so many classes of society, that one would regard a similar result as effectively impossible to obtain. It is not so for the hordes of savages, above all, for those who people New Holland and Van Diemen's Land: here, the population is above all so feeble that the most numerous of the small tribes scarcely number some individuals; and for the most part are less than fifty; here, all difference of conditions of exercise, of food is unknown; with the same needs, with the same resources, all the individuals have the same amount of work to support, the same privations to suffer. This uniformity, which is reproduced in all the details of their existence, which underlines all the epochs of their life, impresses on individuals a character of physical and moral similarity of which one has difficulty in forming a just idea in our social state. Also, I did not hesitate, in spite of the small number of observations that I am going to present on the strength of the inhabitants of Van Diemen's Land, of New Holland, and the same of Timor, to regard these results as much more real [*positif*] than those which I have been able to draw out of the more numerous series of experiments made on European people. I am thus far from wishing to deduce from the dynamometrical observations that I have been able to make on the English and the French, any consequence of the real strength of the two nations in Tables IV and V; I present them only as a term of comparison to bring out more clearly the weaknesses of savage people.... This slightly larger difference for the English seems dependent for the large part on disparities in the health of the individuals of the two nations, the French, about to sail the seas again and the English found, on the contrary, in their homes; it could, thus, be determined by other small, local circumstances of the individuals, that it would be ridiculous to want o draw out some inductions on the relative force of one and the other.

5. PRESSING THOUGHTS

For Peron, his meagre numerical measurements clearly prove that the inhabitants of Timor, New Holland and Van Diemen's Land are of a feeble constitution and much weaker than European men, in spite of accompanying descriptions appearing to indicate otherwise. 'Must one impute this feebleness exclusively to the structure of the societies in which they live, or even to the absence of all social organization among them?' Peron rhetorically asks (1807: 150). He proposes that the cause has social origins rather than being the result of constitutional disabilities. Such an approach is largely fashioned on the ideas that Peron had received during his medical education at the Ecole de Medecine, at that time dominated by the theory of hygiene. Hygiene was an elaborate version of eighteenth-century environmentalism which argued that the complexity of the environment (in its broadest sense, including such features as diet and elasticity of air) determined the degree of civilization available to its inhabitants.

This newly-proved feebleness inherent in the individual savage is not to be mistaken for an inevitable and permanent state of savage society. Despite savage society causing this weakness, judicious education could enable that society to rise to 'higher' stages of civilization. Underlying this is a four-stage theory of societal evolution: hunter/gatherer, pastoral, agricultural, and commercial. Degerando, for example, argues 'If one found some means to transform savage peoples to the condition of herdsmen or husbandmen, one would, no doubt, open before them the surest route that could lead them to the advantage of civilization'. At another point, however, Degerando argues that the creation of a need for trade (with its concomitant sense of property and ownership) would also lead to civilization. It thus seems that it was not necessary to go through all four stages of civilization sequentially. A savage society could be raised directly to the echelons of contemporary European society, implicitly for the advantage of all (Degerando, 1800: 95, 97).

An important feature of Hygiene was its rejection of a deterministic environmentalism in favour of a system that allowed the environment to be modified or manipulated. Such an approach had, in part, been facilitated by the fervent belief of post-Revolutionary France that the power was in one's own hands to change the future. People were not condemned to one form of society, as *ancien regime* environmentalism had had it. Such determinism, together with Rousseau's conception of savage society as the most noble of all, was anathema to the progressive philosophical doctrine of turn-of-the-century France. This aspired to control disease and to improve the health and morality of the people through proper environmental and social reform. A social model of disease underlies this agenda.

The ideal, such 'modern physiologists' as Peron (1807: 422) argued, involved wholesome yet moderate food, climate and exercise. The environments of Timor and Australia provide two opposite extremes to this, and hence can be seen as an *in viva* laboratory to test these hygienist claims. The very fertile, humid and luxurious climate of Timor, with its too easy access to food, promoted 'feebleness and indolence'. Fish, fowl and fruit can be 'procured without work' which results in a country whose residents have 'an apathetic character and a very strong aversion and resolute indifference to work and labour'. Peron seems not to approve of this lifestyle of guitar and flute playing, swimming, chewing betel-nut and sleeping whenever tired. He implies that it lacks morality, that it is too easy. Peron's judgements here, although directed against Timor's inhabitants, is really targeted at Enlightenment *philosophes* who had idealised the tropical life as an idyllic existence epitomizing not only the perfect state of nature, but also the perfect type of society. It is this image that Peron wishes to undermine. 'The high temperature of Timor, its habitual humidity, the indolent life of its inhabitants seem to me to furnish by themselves a satisfactory explanation of [this] particular weakness' (1807: 422-7). In contrast to the rich abundance of Timor, 'Nature seems to have treated the inhabitants of [New Holland and Van Diemen's Land] harshly'. For the

French, the aboriginal diet of frogs, snakes, lizards, caterpillars and tree bark was one which 'the most hideous famine alone could have inspired'. To obtain sustenance, the aborigines are forced to extremes of exercise: 'The savage...indulges over several days in long and arduous hunts, only resting for brief moments when his body has collapsed with fatigue and exhaustion', and when the food is procured, commits the 'excesses and orgies of the half-starved'. Peron argues (1807: 151) that this is nor conducive to the development of physical strength and postulates the almost Lamarckian question of whether the combination of scarce, poor quality food with this form of exercise 'have possibly, if not determined originally, then at least exaggerated this thinness, this excessive emaciation of these men ...?'

Although the 'habitual state of misery and famine' are 'essential causes' of feebleness of the Diemenese, they must 'be considered in their turn as a direct and necessary result of the savage state in which these unfortunate people still pass their lives'. Following the four-stage theory of cultural evolution, Peron argues that 'less slender and less wasted forms' could be achieved by the introduction of the benefits of pastoral activity, the second stage of civilization, where 'the privileges of property come to excite in their breasts a happy emulation'. He paints an idyllic picture of the aborigine at rest in his cabin, with gambolling kangaroos, wombats and emus in the surrounding farmyard and the lyrebirds and black swans gracing the landscape - and the dinner table. Social consequences, such as improvements in the arts and in 'the happy communication of ideas', will arise from this. Peron claims to have refuted 'the eloquent harangues of the sophists', arguing that, since feebleness is due to the scarcity of resources, which, in turn, depends on 'the social condition', 'the progress of civilization would become here the twin sources of vigour and physical perfection' (1807: 152-3).

6. CONCLUSION

Peron's conclusions, and the response to these 'experiments', illustrates the persuasive power of the experimental method and of quantification in the human sciences. At the primary level, Peron and his fellow philosophical travellers give more credence to the dynamometer readings than to any other evidence of strength or endurance in both Europeans and 'savages'. At the secondary level, Peron's conclusions, despite measurements indicating a substantial overlap between races, were not only accepted without question but were also endorsed by his audience, not least amongst whom were members of the Institut de France. Even after Freycinet had found that Peron had misread the loin measurements, Peron's results were not taken to be potentially spurious: they were merely adjusted. The interposing of the physical instrument, the dynamometer, between the observer and the observed led to a confident belief that the ensuing measurements had eliminated any subjective element. The errors of measurement were rationalized as merely mechanical mistakes. Reading dials and recording data were understood to be purely passive activities, untarnished by any possible bias or expectations on the part of the scientists. Adjustments made after the acknowledgement of error were equally assumed to be free from influence. It became acceptable to privilege some results above others, as is apparent in the quotation where Peron 'explains away' the discrepancies between the French and English readings.

One could argue, hypothetically, that the difference between these two groups of the French and the English is only slight and Peron's explanation is reasonable while that between, say, the English and the Diemenese, is, as Peron claims, more 'real'. However, it is clear that in Peron's case he has a strong belief that his numbers reflect absolute strength. It is inconsistent, on the one hand, to provide a special pleading to the French because there is an apparent explanation, and on the other hand, to deny outright that the Diemenese's low reading is perhaps also due to a transient cause. Whilst allowing that factors of lifestyle can distinguish between the French and the English, Peron, by contrast, is then arguing that the Diemenese are inherently weak.

The extent of the faith in the objectivity of the dynamometrical numbers is revealed in Peron's comments at Île de France. Here, the dynamometer was left with Chapotin, the government medical officer, in the hope, Peron writes, 'that in his hands it would one day furnish the most valuable results'. Peron had had the opportunity to test a variety of different races in his voyage to the Southern Hemisphere (though it appears that he only took eight-five non-European measurements). He hoped that Chapotin would augment his results by studying such groups as 'the wild Yolof, the haughty Malagasy, the huge and bellicose Kaffir, the sturdy Mozambique and the more peaceable inhabitants of the Sunda Islands and the Moluccas, as well as large numbers of those indolent, feeble and soft peoples who come from the regions of the Indus and the Ganges' (1807: 150). Peron's suggestions for these measurements are in keeping with the spirit of scientific investigation in their advocacy of a comprehensive comparative analysis of races based on the results of the dynamometer.

The report of Coulomb and Labillardiere to the Institut approved both the approach and the conclusions of Peron. It speaks of 'proper observations' establishing 'with much accuracy' that the Diemenese lifestyle is characteristic of 'non-social man'. It argued that the lack of willingness that the Diemenese showed for being measured strengthened Peron's conclusions. The referees also agreed that much more research would be required to 'know the actual strength of civilized peoples'. Coulomb and Labillardiere mention some of the 'very interesting historical details...of the unhappy life of savages'. They conclude (1804: 235-6): 'We again observe that the author is the first to have carried the dynamometer to the Southern Seas.' The endorsement of Peron's conclusions, primarily that strength is directly proportional to civilization, reveals an explicit belief in the validity of the numerical results for the measurement of strength. The results of the experiments were seen to provide scientific legitimacy for both a cultural and a racial hierarchy, since the two were often conflated. Indeed, the dethronement of the Noble Savage - or at least, his 'robust majesty' (Peron, 1800: 183) - would prove to be more important politically than if the dynamometer results had 'proved' such mythical strength.

On the basis of Peron's twenty-four measurements, the whole autochthonous population of Australia was condemned to the category of feebleness. This signals the transition to the nineteenth-century approach to the study of man based on notions of scientific objectivity and impartiality. Degerando's advice of not treating an individual as a representative of the whole race, a belief rooted in eighteenth-century liberalism, was overthrown for a perspective that rendered the savage as no more than a variety of animal, where one measurement acted as a sort of type specimen, regarded as actually true for the whole race. In this context, one should note parenthetically that it was the slave trade at Île de France - 'the activities of our Europeans' (Peron, 1807: 151) - which brought these many diverse races to the islands.

Finally, one might speculate about the conclusions that the Diemenese subjects might have reached. Peron attributes motives to the Diemenese to explain their behaviour and this shows a degree of sensitivity to the impact that the French presence had had on that community as well as a perceptive insight into the ultimate fate of the natives (see Latour, this volume; Horner, 1987; Plomley, 1983). But we cannot know that such an interpretation is correct, despite its beguiling reasonableness. The Diemenese's reluctance to measure their loin strength shows an understandable uncertainty. Surrounded by arm-wrestling, gun-wielding, Marseille-singing, trick-performing Frenchmen, what could the natives think of the experiment?

ACKNOWLEDGEMENTS

I thank Homer Le Grand and Bruno Latour for comments on an earlier version of this paper. I am grateful to the Archives de France and the Museum d'Histoire Naturelle at le Havre for permission to quote from their holdings.

NOTES

1. The writings of Montesquieu and Rousseau are representative of this approach. The 'true', 'natural' state of human society is imagined and from this a critique of contemporary society evolves.
2. Baudin (1974: 576). Some ninety years later, much praise was given to A.C. Haddon (1901) for performing the 'first' experiment to test the strength of Torres Strait Islanders.
3. Archives Nationales, AJ 15 569, feuille 367.
4. For a more general account of the anthropological instructions and encounters in Australia see Jones (1988) and Hughes (1988a; 1988b). A detailed account of Baudin's expedition is found in Horner (1987). The anthropological and zoological artwork is reproduced in Bonnemains, Forsyth & Smith (1988).
5. Peron (1802: 88-9) provides this synopsis at the side of the paper to describe the events: Particulars of our friendly conduct towards the native. We were not able to triumph over their distrust. Singular precautions against any surprise from us. Cause of their disquiet. They become a little calmer. They put down their arms again. They no longer wish to reply to me. They become more venturesome but fear the gun. They torment us in order to engage us in fire at birds. One of them threatened me. He brandished his spear at me. Adroit way I adopted to intimidate him. (Peron laughs, appears friendly, points to his gun and says mata [dead].) He lays down his weapon. Another wishes to tear off one of my earrings. He misses his chance.

REFERENCES

Baudin, N. (1974). *Journal of Baudin, 1800-1803.* Trans. C. Cornell, Adelaide: Library Board of South Australia.

Bonnemains, J., Forsyth, E. & Smith, B. (1988). *Baudin in Australian waters. The artwork of the French voyage of discovery to the southern lands, 1800-1804.* Melbourne: Oxford University Press and Australian Academy of Science.

Coulomb, C.A. (1800). Resultat de plusieurs experiences destinees a determiner la quantite d'action que les hommes peuvent fournir par leur travail journalier, suivant les differentes manieres dont ils emploient leurs forces. *Memoires de mathematiques et de physique de l'Institut de France*, 2 (an VIII), 380-428.

Cuvier, G. (1800). Note instructivesur les recherches a faire aux differences anatomiques des diverses races de l'homme. In *Aux origines de l'anthropologie francaise: Les Memoires de la Societe des Observateurs de l'Homme en l'an VIII*, ed. J. Copans & J. Jamin, pp. 171-6. Paris: Le Sycamore, 1978.

Degerando, J.M. (1800). *Considerations sur les diverse methodes a suivre dans l'observation des peuples sauvages*. Trans. F.T.C. Moore, *The observation of savage peoples*. Berkeley: University of California Press, 1969.

Haddon, A.C. (1901). *Reports of the Cambridge Anthropological Expedition to Torres Straits. Vol.2: Physiology and psychology*. Cambridge: Cambridge University Press.

Horner, F. (1987). *The French reconnaissance: Baudin in Australia 1801-1803*.Melbourne: Melbourne University Press.

Hughes, M.J. (1988a). Philosophical travellers at the ends of the earth. In *Australian science in the making*, ed. R.W. Home. pp. 23-44. Sydney: Cambridge University Press and the Australian Academy of Science.

Hughes, M.J. (1988b). Tall tales or true stories? Baudin, Peron and the Tasmanians, 1802. In *Nature in its greatest extent: Western science in the Pacific*, ed. R. MacLeod & P. Rehbock, pp. 65-86. Honolulu: University of Hawaii Press.

Labillardiere, J. & Coulomb, C.A. (1804). [Report on Peron]. *Academie des Sciences proces-verbaux*, 3 Thermidor an 13, 235-6.

Peron, F. (1799). Observations sur l'anthropologie,ou l'histoire naturelle de l'homme, la necessite de s'occuper de l'avancement de cette science, et l'importance de l'admission sur la flotte du Captain Baudin d'un ou plusieurs naturalistes, specialement charges des recherches a faire sur ce sujet. Paris, an VIII. Reprinted in J. Copans & J. Jamin (1978), pp. 177-85.

Peron, F. (1802). Maria Island: Anthropological Observations. Printed in and trans. by N.J.B. Plomley, *The Baudin expedition and the Tasmanian Aborigines, 1802*, pp. 82-95. Hobart: Blubberhead Press, 1983.

Peron, F. (1806). Experiments on the physical strength of the savage people of Van Diemen's Land and New Holland and of the inhabitants of Timor. Trans. & rept. in Plomley (1983), pp. 145-59.

Peron, F. (1807). *A voyage of discovery to the Southern Hemisphere performed by order of the Emperor Napoleon, during the years 1801, 1802, 1803, and 1804*. Rept. Melbourne: Marsh Walsh Publishing, 1975.

Peron, F. & Freycinet, L.C.D. de. (1816). *Voyage de decouvertes aux terres australes, execute par l'ordre de Sa Majeste L'Empereur et Roi, sur les corvettes le Geographe, le Naturaliste, et la goelette le Casaurina, pendant les annees 1800, 1801, 1802, 1803 et 1804; (Historique)*. Paris: Imprimerie imperiale.

Regnier, E. (1798). Description et usage du dynamometer, pour connaitre et comparer la force relative des hommes, celle des chevaux et de toutes les betes de trait; enfin pour juger la resistance des machines et estimer les puissances motrices qu'on veut y appliquer. *Journal de l'Ecole Polytechnique*, 2, 160-78.

University of Melbourne

JAN GOLINSKI

HUMPHRY DAVY AND 'THE LEVER OF EXPERIMENT'

The love of discovery, and of intellectual acquirement, may indeed be called an almost instinctive faculty of the mind; but still a combination of circumstances is required to call it into useful exertion. The truly insulated individual can effect little or nothing by his unassisted efforts. It is from minds nourishing their strength in solitude, and exerting that strength in society, that the most important truths have proceeded. Humphry Davy, (1839-40, **8**: 321)

1. INTRODUCTION

How do people become convinced by experiments? Why should anyone draw any conclusions from them? And why should they draw particular conclusions? These questions have emerged as central ones in recent work in the history and sociology of science. Their prominence in contemporary discussion of scientific practice marks the degree to which we have departd from a naive philosophical view of the nature of science. From that point of view, experiments are taken to represent nature just as it is and they simply communicate natural facts. Not to be convinced by them would be to violate the basic methodological rules of the scientific enterprise.

Scrutiny of actual scientific practice has shown that this naive view is wrong. Scientists do not believe all experiments unconditionally and it would be a foolish philosopher who would recommend them to do so. Experiments sometimes contradict one another, so that they cannot all be believed: some selection of 'reliable' or 'credible' results is necessary. Furthermore, experiments by themselves cannot produce the theoretical structures of science; they require supplemental interpretation. Recent research in the sociology of scientific knowledge works from the basic proposition that experimental evidence is always *too much* to fit within any particular theoretical construction and always *too little* to decide of itself between any number of plausible theoretical views.

The effect of accepting this is to direct attention to the context in which experiments are performed and used. The question of how scientists (or non-scientists) become convinced by experiments is opened

H. E. Le Grand (ed.), Experimental Inquiries, 99–136.

to investigation and the answer is sought in the contexts in which experimental work finds its meaning. In particular, recent historical and sociological studies have concentrated on how experiments are used to sustain interpretive claims about the natural world, on the ways in which they are used in communication and persuasion.[1] Experiments are given meaning by being mobilized in support of certain claims; *i.e.*, by being described in particular forms of discourse and deployed in efforts to convince certain audiences. They acquire their significance in a context of *rhetoric*, a context that includes language and social relations as essential elements.

I adopt a phrase used by my subject, the English chemist and natural philosopher Sir Humphry Davy (1778-1829),[2] to describe this situation. Davy used the phrase 'the lever of experiment' (1839-40, 8: 318) to refer to the power of certain significant manipulations to stir nature from her hiding-place and thereby force her to reveal herself. His usage in some respects foreshadows Bruno Latour's description (this volume) of 'the leverage point of experiments'. In Latour's view, experiments connect together material objects, semiotic or rhetorical constructions, and social actors. They put to the test the links between these heterogeneous entities, sometimes forcing them apart or making them re-align in new patterns.

I shall use Davy's phrase in a way which is different from his own usage or from Latour's, namely to draw attention to the social leverage exerted by experiments. Like Latour, I accept that manipulations of material objects can only move opinion if they are embodied in appropriate linguistic forms and communicated *via* functional social relations. In other words, the lever of experiment turns upon the fulcrum of the relations between experimenter and audience. However, I shall not follow Latour in speculating on the ontological dimensions of this process, or on the roles of non-human 'actants' therein. Instead, I limit myself to describing the observable effects of the lever of experiment: its manifest power of securing the acceptance of beliefs.

I shall describe two episodes from Davy's career to discern how he used the lever of experiment in each case. The first episode centres on Davy's first Bakerian Lecture, delivered to the Royal Society in 1806. In that lecture and in work connected with it, Davy converted the Voltaic pile (a primitive electrical battery) from an 'experiment' into an 'instrument' with which reliable chemical analyses could be performed. This transformation will be shown to have been a social as well as a technical achievement; it was one that depended as much on the mobilization of certain rhetorical forms, and on the establishment of a relationship of authority over an audience, as on material manipulations in the laboratory. Davy had to engage in a strenuous defence of the efficacy of the Voltaic pile as an instrument of analysis before his discoveries with it could win recognition. Having established its instrumental efficacy by fending off alternative interpretations of the pile's action, Davy could use it to make his famous discoveries of the elements sodium and potassium. These discoveries gained wide acceptance but still required further defensive actions on Davy's behalf to defeat occasional challenges to their existence or their status as elements.

The second episode concerns Davy's announcement of the elementary nature of chlorine in 1810. In the ensuing, prolonged debates, his claim was subjected to severe scrutiny by other members of the chemical community and his experimental evidence was brought into question. To win his argument, Davy deployed various rhetorical techniques and attempted again to enlist the support of public (non-specialist) audiences for his claims. These tactics had only limited success. Some of Davy's opponents remained stubbornly unconvinced; most of those that were won over were finally persuaded not by *his* work but by the French discovery of what appeared to be an analogous element, iodine. In this incident, the lever of experiment had not proved such an effective mover of opinion.

2. THE VOLTAIC PILE: THE MAKING OF AN INSTRUMENT

In pliant obsequiousness to theory, the cathode jar was almost full of gas; the anode jar was half full: I brought this to Enrico's attention, giving myself as much importance as I could, and trying to awaken the suspicion that, I won't say electrolysis, but its application as the confirmation of the law of definite proportions, was my invention, the fruit of patient experiments conducted secretly in my room. But Enrico was in a bad mood and doubted everything. 'Who says that it's actually hydrogen and oxygen?' he said to me rudely. 'And what if there's chlorine? Didn't you put in salt?' (Primo Levi, *The Periodic Table)*

Davy made the Voltaic pile into an instrument of experimental research by wresting it from its original context in late eighteenth-century natural philosophy. Galvanism, like Mesmerism, was initially subject to the politically-polarized interpretations of the 1790s, the decade of 'the end of the Enlightenment'. It was seen by the progressive proponents of enlightenment natural philosophy as a manifestation of an imponderable fluid similar if not identical to aether, the underlying cause of heat, light, electricity and even nervous action. Davy's early associate and mentor Thomas Beddoes hailed the discovery of the Voltaic pile as significant for this reason. It would, he wrote, form 'the origin of a new and more subtle chemistry; in which etherial fluids ... will make the principle figure'.[3]

From the other side of the political spectrum, the anonymous author of *The Sceptic* (1800) grouped galvanism with oxygen, Mesmerism, and revolutionary hysteria itself, as phantom spirits that caused chaos and delusion. The galvanic spirit, this author wrote, 'fled into Italy, and entered into the body of a Frog, which professor Galvani was dressing for his Wife's supper' (1800: 6). The phantom was destined to remain invisible and completely useless.

Davy's early notebooks, compiled mostly during his residence in the West Country before he moved to a professorship at the Royal Institution (R.I.), show his interest in galvanism to have originated in aspirations very like those expressed by Beddoes. Galvanic phenomena promised to open a window into the nerv us system. Galvanism seemed to Davy to connect the organic and inorganic, the material and mental, realms.[4] Hence in July 1800, Robert Southey wrote to him that the Voltaic pile was important because it implied that, 'the galvanic fluid(,) stimulated to motion,...is the same as the nervous fluid, and your systems will prove true at last'.[5]

Thus Davy shared the enthusiasm for galvanic investigations which swept Europe in the wake of Luigi Galvani's description of his original experiment in 1791. Galvani's claim to have produced motion in the limb of a frog by applying a metallic couple to connect nerve and muscle, generated intense controversy. Dispute focused on the questions of the origin and the nature of the galvanic fluid: did it originate in the frog's nerves, or in the metals applied to the limb? Was the fluid the same as the electricity commonly produced from electrical machines? The discovery of the Voltaic pile in the spring of 1800 refocused these disputes, but did not finally resolve them (Kipnis, 1987). The pile showed that galvanic fluid could be produced from combinations of inorganic materials, but it did not kill off research into animal sources of electricity: Davy himself continued to take an interest in this field throughout his career (1839-40, **2**: 221-8; **6**: 359-64). Nor did the pile decide the question of the identity of the galvanic fluid with common electricity. Few in fact accepted this identity, even in 1800 (Kipnis, 1987: 125-31, 136; Davy, 1839-40, **2**: 201, 205).

The importance of the Voltaic pile for Davy hinged upon its use by William Nicholson and Anthony Carlisle in July 1800 to effect what they claimed was the electrical decomposition of water.[6] They described how they had inserted wires from the opposite poles of the battery into water. This apparently produced hydrogen gas from one electrode, and oxygen, which formed a compound with the metal of the other electrode. Davy immediately accepted this result, realizing that it implied the possibility of using the Voltaic pile as an analytical instrument for resolving hitherto undecompounded chemical substances. On 6 August 1800, he wrote that Nicholson and Carlisle's result had 'placed such [a] wonderful and important instrument of Analysis in my power' (Davy MSS, box 20b, 2).

In 1800 Davy saw the Voltaic pile as potentially a powerful instrument of analysis. However, in order that it should realize this potential, it was clearly necessary for it to become widely accepted as effective and reliable. In sociologists' terms, the pile had to be 'black-boxed'; *i.e.* transformed from a subject of experimental investigation into an unproblematic tool for further research. The integrity and efficacy of the pile as an analytical instrument depended crucially upon the acceptance among chemical researchers of Nicholson and Carlisle's result.[7] Unless it were agreed that the pile really did decompose water, into *just* hydrogen and oxygen, it could not function as an instrument of analysis at all. Davy faced an uphill task, for in 1800 there was no general acceptance of the claimed electrical decomposition of water.

The source of the confusion about the mode of action of the pile on water was that the apparatus was relatively easily replicated. To construct a pile, one needed only metal wires, pieces of damp cloth or cardboard and metal discs (silver half-crowns were popularly used and discs of zinc could readily be obtained).[8] Replication, to the extent of producing some kind of action upon water, was easily and rapidly achieved. There was little agreement however as to what was going on. The result was a plethora of papers in periodicals such as *Nicholson's Journal* and *Philosophical Magazine*, suggesting a variety of interpretations of the pile's action. Contributions to the debate appeared from a wide range of geographical and social locations; William Cruikshank and Col. Henry Haldane from London, William Henry from Manchester, Charles Sylvester from Sheffield, William Peel from Cambridge, D. Gardner (Lecturer in Chemistry at the City Dispensary, London), Joseph Priestley from Pennsylvania, and other experimenters from Scotland, France, Italy and Germany.[9]

Some of these experimenters attacked the assertion that water was decomposed by the action of the pile. Priestley published a paper (1802) denying the Voltaic decomposition of water, as part of his rearguard attack on the whole of the new French theory of chemistry. The 'dephlogisticated air' produced at the zinc wire came from air dissolved in the water, Priestley asserted; it was not evolved if the water was covered with a layer of oil, or if it had been 'exhausted' in the receiver of a vacuum pump. He also claimed support from several other contributors to *Nicholson's* for the view that the Voltaic pile actually demolished the 'chimerical hypothesis' of Lavoisier.

The fact that the supposedly constituent gases were produced at different electrodes, which might be widely separated, was a serious difficulty for Nicholson and Carlisle's claim. William Cruikshank, writing in September 1800, basically accepted their result, but noted hesitantly, '...how this can be effected, is by no means so easily explained' (1800-01: 257). For others, the absence of a plausible explanation of this phenomenon was an insuperable obstacle to accepting that the pile acted analytically. 'An experimentalist', writing in *Nicholson's* asked how it was possible for a component particle of water, 'insensibly to hurry through the water for a distance of six inches or more, and there to make its appearance in the character of gas'. He concluded (1800-01: 515):

> It appears to me therefore that the separation of the two gases from the connection with the pile of Volta, is...unexplained according to the doctrine of the decomposition of water. ...I do not see how, in strict philosophy, we are warranted in saying more of the experiment with the pile, than that one kind of electricity and water produce one kind of gas, and the other another gas. Is it not an assumption in this experiment to say, that the bases of the gases are the component parts of water?

This alternative explanation for the pile's action gained greater credence during 1801, as the work of Johann Wilhelm Ritter was reported from Germany ('Dr. G.M.', 1800-01). Ritter took the separate production of hydrogen and oxygen as conclusive proof that they were not the components of water. A more likely explanation of what was occurring, he believed, was that oxygen was produced from water plus positive electricity, and hydrogen from water plus negative electricity. In Ritter's view, water was thus shown to be what he had always maintained it to be, an element.

The Voltaic decomposition of water did have its defenders. Charles Sylvester mounted a strong attack (1804) on those who claimed (as Ritter had) that there were two types of galvanic electricity. This was a more arbitrary assumption, Sylvester asserted, than to accept that the two components of water were produced separately. But, Davy's papers from this period evaded the problems of understanding the effect of the pile upon water. His investigations of its mode of action, published in about a dozen papers between 1800 and 1803, simply assumed its efficacy as an analytical instrument (1839-40, **2**: 139-228; *cf.* Russell, 1959; 1963). This was a perilous assumption in view of what was clearly not a 'closed' debate. Davy's research could be undercut by renewed challenges to his understanding of the way the pile operated on water. In 1805 and 1806, it looked as if this were just what would happen. From a number of sources came repeated claims that acids (or occasionally alkalis) were produced *de novo*, from the action of the Voltaic pile on water. If these claims had been accepted, the effect would have been to undermine the status of the pile as an instrument of chemical analysis.

William Peel wrote (1805) to Alexander Tilloch that he had produced muriates (*i.e.*, chlorides) of soda and potash, by applying the Voltaic current to pure water (Peel, 1805). Tilloch expressed reservations about Peel's techniques but his results did seem to confirm those reported by the Italian experimenter Pacchiani, who had described producing muriatic and oxymuriatic acids (*i.e.*, hydrochloric acid and chlorine) from galvanic action upon water. Other investigators, Cuthbertson (1806) in London, Brugnatelli and Alemani in Italy, confirmed that muriatic acid was indeed produced, no matter how stringent the precautions taken to use only distilled water and to avoid any contamination by organic substances.[10] Defenders of the analytical power of the Voltaic pile were therefore given cause to realize by the end of 1806 that the efficacy of their instrument was severely threatened. If these results were accepted, the pile was not acting simply to analyze water into oxygen and hydrogen. Somehow acids (muriatic, oxymuriatic, and nitric) were being produced by its action, perhaps, as Sylvester speculated (1806: 50-2), by oxidizing hydrogen to various degrees.[11]

This was the situation that Davy confronted as he delivered his first Bakerian Lecture to the Royal Society on 20 November 1806. To impose order on this confusion and protect the power of his apparatus, he had first to establish his authority to address a wide audience and then to convince it of his command of the instrument. Next he had to muster his arguments, to reduce the freedom for alternative interpretations and to marginalize those who upheld them. The refinements of experimental technique that he introduced were directed towards blocking off various possible alternative views of the pile's action. They also had the effect of making replication of the experiments more difficult, thereby securing Davy's claims from challenge by those who did not share his resources of equipment and skill.

The authority over his audience and over his instrument that Davy deployed had been acquired during his public lectures at the R.I. during the previous five years. There he had cultivated a repertoire of experimental demonstrations adapted to a situation very different from the provincial setting of his early work with Beddoes. By moving to London in 1801 from Beddoes's Pneumatic Institution in Bristol, Davy had been translated, as a friend wrote, from 'the stormy Parnassus' to 'the mild and unvarying temperature of the central grotto of science'.[12]

At the R.I. Davy addressed hundreds of members of London's fashionable elite, along with leading gentlemanly specialists in experimental science, and his lectures were widely reported in cultural periodicals and the general press.[13] A commentator on his very first lecture series, on galvanism, described those attending as, 'Men of the first rank and talent, ...blue-stockings and women of fashion, the old and the young' (quoted in Davy, 1839-40, **1**: 88). Thomas Carlyle aptly joked that Davy had made the R.I. into 'a kind of sublime *Mechanics' Institute* for the upper classes' (quoted in Foote, 1952: 7).

The prominence of women among his audiences was a feature which many commentators noted, and which Davy himself welcomed. He regarded the presence of women at cultural events as a sign of social refinement and progress (1839-40, **8**: 353-5): 'The standard of the consideration and importance of females in society is, I believe, likewise the standard of civilisation'. His view of the position of women in intellectual life gave them an important but limited role among his audience: they were expected to absorb and transmit science as part of a general cultural education but not to attempt to penetrate into the realm of the specialist. This view had been earlier proposed by such scientific educators as Priestley and Adam Walker, and it subsequently inspired Davy's pupil Jane Marcet to compose her *Conversations on Chemistry*.[14] Women were to be admitted as part of an admiring audience for 'popular' science, but detailed or technical knowledge should remain a masculine preserve.

That Davy's lectures were masculine science, presented to an audience that was, to a significant extent, feminine, was one important dimension of their construction. But, male and female auditors alike found themselves seduced by his mode of delivery and captivated by his forceful presentation of his own personality, and of the powers of nature under his control. His carefully cultivated oratorical talents and attractive personal appearance were mobilized to produce this effect, as was his choice of convenient and spectacular experimental demonstrations.[15] Through these means, Davy appeared in the guise of the scientific 'genius' reflected in his own deliberations on that character in his private notebooks.[16] Most observers of his lectures would have agreed with the historian Harriet Martineau (1849-50, **1**: 594), that Davy 'presented most strongly to the popular observation the attributes of genius'.

Among the attributes of genius, as the term was understood by contemporaries, was a special insight into the powers of nature and an ability to command them at will. In his lectures Davy made much of these powers, and of recent discoveries which gave promise of their more effective utilization. A prime example occurred in his 'Discourse Introductory' delivered at the R.I. and published a few months later (1839-40, **2**: 311-26). In this lecture Davy echoed many of the traditional justificatory themes, describing especially the extensive utility of the subject. He also added a topical emphasis on natural powers and the instruments that might allow them to be controlled. 'New instruments and powers of investigation' were invoked to show that chemistry, as opposed to those sciences which only studied surface appearances, could penetrate into matter itself. It would be to chemistry therefore that natural philosophers would have to turn to understand the fundamental internal processes which distinguished living beings (**2**: 312-4). This was an area in which Davy showed significant interest in the course of his career and it was one in which he saw imminent prospects of results from the new instrumental powers now at his control. Precisely because they offered the promise of knowledge of the fundamental processes of life, including human life, these new powers gave a god-like status to the philosopher who wielded them. They enabled man, Davy explained, 'to interrogate nature with power, not simply as a scholar, passive and seeking only to understand her operations, but rather as a master, active with his own instruments' (**2**: 319). Specifying these powers, he reviewed recent discoveries in natural philosophy (**2**: 321):

> The composition of the atmosphere, and the properties of the gases, have been ascertained; the phaenomena of electricity have been developed; the lightnings have been taken from the clouds; and lastly a new influence [galvanism] has been discovered, which has enabled men to produce from combinations of dead matter effects which were formerly occasioned only by animal organs.

Many of these powers were shown by Davy in his lectures at the R.I. Under his direction, the forces of nature were shown to be, as he put it, 'according to circumstances, instruments of comfort and enjoyment, or of terror and destruction' (**2**: 318). In the lectures, they were demonstrated primarily in order to manifest their efficacy as tools of investigation, by showing that they could be mastered by the experimenter to yield useful knowledge. The rhetorical force of such demonstrations could be considerable in establishing acceptance among his audience that the forces of nature were indeed Davy's to command. He recognized the peculiar potency of experiments to convey this message, having written in one of his early notebooks that 'the language of exp[erimen]t is universally intelligible and the truths it conveys can never be forgotten' (Davy MSS, box 13c, 94).

To teach this lesson, it was essential to choose appropriate experiments: ones which would enhance the apparent mastery of the experimenter; not ones which produced disorder or confusion. Davy made more than one attempt in his early days at the R.I. to demonstrate the effects of breathing nitrous oxide, one subject of research he had conducted at Beddoes's Pneumatic Institution. The results were striking in a quite undesirable way. At a lecture in June 1801, he offered the gas to those among his audience who wanted to breathe it. The outcome was unexpectedly comical and disruptive: a Mr. Underwood 'lost all sense to everything else, and the breathing-bag could only be taken from him at last by force'. Perhaps to avoid a repetition of these unbecoming scenes, the experiments were continued a few days later at a specially convened private meeting. Respiration of nitrous oxide appears to have dropped out of the repertoire of public experiments after this, perhaps reflecting a desire to avoid ridicule and disorder. The publication of James Gillray's well-known satirical cartoon, 'New discoveries in Pneumaticks!' in May 1802 may well have also been a factor.[17]

Ridicule of the lecturer was certainly not the outcome Davy sought from his performances. His aim was to control the situation in the theatre so as to give a striking demonstration of the power of his 'instruments' and to co-opt that power to enhance the credibility of the experimenter. Galvanic phenomena produced by the Voltaic pile yielded a fund of such demonstrations, as Davy had shown in his early lectures on this subject. The pile produced shocks, sparks 'of a dazzling brightness' and loud noises. The sparks would be different colours (white, purple, yellow, red or blue), according to which metal composed the electrodes. Party-tricks were shown: gunpowder and mercury fulminate was ignited by the galvanic fluid; letters and figures were drawn on gold-leaf by burning off the metal by electricity (1839-40, **2**: 211-3). At the same time as he showed its spectacular and dramatic effects, he stressed that the visible powers of the pile were as nothing compared with those which could not be seen: 'It is not by flashes, by explosions, and sparks, but by quiet, gradual, and almost unperceived operations, that it produces its greatest effects' (1839-40, **8**: 171). Those who attended the lectures were given forceful visual and verbal evidence for Davy's assertion that Volta had provided 'a key which promises to lay open some of the most mysterious recesses of nature' (1839-40, **8**: 282).

By showing the efficacy of such natural forces as galvanism in his lectures, Davy simultaneously demonstrated his power reliably to command them through the instruments at his disposal. His possession of such instruments was projected as an integral part of his presentation of himself to his audience as an experimental philosopher. In the rhetorical construction of his lectures, the power of the instruments and of the experimenter buttressed one another. It was not just that the efficacy of his instruments validated the actions of the experimenter: the personal genius of the experimenter guaranteed the strength and reliability of his instruments.

The authority over his instruments that Davy had acquired in the eyes of his audiences at the R.I. was a substantial asset to him in his 1806 Bakerian Lecture. The audience which Davy addressed at the Royal Society was different in composition from that at the R.I.: it contained a higher proportion of men knowledgeable about science, and no women. But, it remained a public audience, rather than one composed exclusively of specialists, particularly when the 'penumbral audience' of those who read reports in newspapers and general periodicals is taken into account. Another difference from Davy's previous lectures was that the discourse was entirely verbal, and, by the time of publication, written rather than spoken. No accompanying experimental demonstrations were presented. Literary rather than oratorical techniques were required and the intrinsic persuasive force of displays of natural powers could not be directly invoked. He had to develop new rhetorical methods to both reinforce his authority with this new audience and re-establish the perceived efficacy and security of his prime instrument of research.

The first section of the lecture is arguably the crucial one from this point of view. In a lengthy series of experiments, Davy sought retrospectively to secure the basis of his electrochemical work: the belief that the Voltaic pile was an instrument of analysis which decomposed water into hydrogen and oxygen only. The order in which these experiments were described was probably not that in which they had actually been performed, but one constructed to persuade the reader to accept the integrity of Davy's instrument.

As a rhetorical construction, the first section of the lecture is very effective. He surveyed the alternative claims about the action of the pile, including the latest experiments by Charles Sylvester reported in

Nicholson's only the previous August, and tried to give the impression of a field of enquiry beset with a confused variety of interpretations. He announced what his own conclusion would be: that all the acids and alkalis reported to have been generated in electrolysis were artifacts produced by contamination. He then described experiments to support his view. His apparatus consisted of a large battery, tiny agate cups, pure platinum electrodes, repeatedly distilled water, and a piece of amianthus (purified asbestos) to connect the two separate cups in which the electrodes were placed.[18]

With this arrangement he told how at first he produced both acid and alkali, the former from the positive, the latter from the negative, electrode. However, after repeated operations in the same agate cups, he succeeded in eliminating the production of acid and greatly reducing that of alkali. He concluded that saline matter in the composition of the agate cups, or in 'intimate adhesion in its pores', was responsible for producing the acid and probably some of the alkali. Attention was then directed at what he said was the only other possible source for the alkali: the water itself. He described experiments conducted using pure gold cups, which initially yielded small quantities of alkali at the negative electrode. This quantity of alkali was not increased however long the operation was continued. By slowly evaporating the distilled water he was using, he established that saline matter had been carried over in the original distillation. This matter, and the alkali to which it apparently gave rise under electrolysis, could be eliminated by especially slow distillation of the water (1839-40, 5: 5-8).

Further experiments were then described which showed that alkalis (as well as acids) could be produced from the vessels in which electrolysis was conducted: wax cups produced potash and soda, resin ones yielded potash, and so on. Glass itself yielded soda, as could be seen if a small piece of glass were added to a gold cup in which a negative electrode was operating. Finally, Davy pointed to the presence of dissolved nitrogen in the water as a plausible source for the appearance of acids and alkalis. He cited Priestley as authority for the view that 'nascent' oxygen would co-exist in solution at the positive pole with nitrogen, forming an acid, whereas nascent hydrogen would expel any dissolved nitrogen from the water around the negative pole. This suggestion was apparently confirmed by experiments performed under reduced atmospheric pressure, or with another gas replacing the atmosphere over the cups of water. In these circumstances, Davy claimed, neither acid nor alkali was produced in detectable quantities

(1839-40, 5: 9-12). All these experimental reports were mustered to support his overall conclusion (5: 9):

> To detail any more operations of this kind will be unnecessary; all the facts prove that the fixed alkali [and the acid are] not *generated*, but *evolved*, either from the solid materials employed, or from saline matter in the water.

By driving a wedge between 'generation' and 'evolution' of products, Davy categorized the acid and alkali as artifacts produced by contaminants, while oxygen and hydrogen were genuine products of chemical analysis. Having established the fundamental efficacy of the pile as an analytical instrument, he went on to *use* it as such and to investigate its mode of action. These later parts of the Bakerian Lecture have absorbed the attention of a number of historians who have puzzled over the discrepancy between the metallic-contact theory of the pile's action that Davy advanced there and the chemical theory proposed in his earlier papers (Russell, 1959; 1963; Ziemacki, 1974: 235-45). This analysis however overlooks the importance of the first section of the lecture, which was the linchpin about which the whole turned. If Davy had not responded vigorously and resourcefully to those threatening to undermine the status of the pile as an instrument of analysis, he would not have been able to gain acceptance of its use as such and the investigation of its mode of action would have been pointless. The instrumental power of the pile was not purely natural; it had to be consolidated by social and rhetorical means.

In written form, the Bakerian Lecture exhibits a series of strategic moves toward solving the problem of the sources of the reported acids and alkalis. Davy appears to encircle the troublesome substances, progressively cutting off their potential hiding-places. The 'contaminants' are never removed entirely, and each claimed identification of their sources is potentially open to challenge but the reader is induced to acquiesce in Davy's assurance that they *could* be eliminated. Hence, one is led to accept his conclusion: 'It seems evident then that water, chemically pure, is decomposed by electricity into gaseous matter alone, into oxygen and hydrogen.'

Davy's text operates in a similarly strategic manner against the potential scepticism of its readers. His apparently exhaustive elimination of possibilities operates as a kind of 'buried dialogue', anticipating and answering the objections of a sceptical reader. The length of the procedure as described and the repeated emphasis on refinements of technique and sensitivity of tests convey an image of Davy as an extraordinarily careful and skilled experimenter. The technical refinements also made experiments in this field harder, raising the stakes in terms of the necessary level of equipment and skills for making contributions in the future. Investigators who lacked these resources would no longer need to be taken seriously.

Most reviewers of the Bakerian Lecture readily accepted its conclusions. Some also commented on its rhetorical effectiveness. Even those who confessed they could not follow the details of the argument said they were convinced by its overall drift.[19] Brougham pointed to the effect of the detailed argument of the first section of the lecture in supporting Davy's other experimental claims: 'The examination of [this enquiry] gives us an irresistible disposition to confide in all the other processes of the author, which he passes over more hastily, or only refers to in general terms' (Brougham, 1807-08: 398).

As Brougham noted, Davy had augmented his future credibility by virtue of the persuasiveness of his arguments on this occasion. By establishing the efficacy of the pile as an instrument of analysis, he had armed himself with the means of producing new knowledge of the chemical composition of matter. And, by persuading his audience that he possessed a reliable instrument, he had gained at least provisional acceptance of the knowledge it would yield. Thus the foundation was laid for the discoveries of two new metals, announced by Davy in his second Bakerian Lecture in November 1807 (1839-40, 5: 57-101). He described how his expectation that the pile could be used to analyze previously undecompounded substances had borne fruit. A current, passed through a sample of fused potash or soda, yielded globules of the new metals, which were named potassium and sodium. The pile was taken to have produced these metallic elements by analyzing their oxides, potash and soda. The 'discoveries' were very widely, if not universally, accepted. Some English reviewers hailed them as the greatest contributions to science in recent times.[20]

It is worth pausing to consider how these 'discoveries' had been achieved. They were produced by utilizing the pile as an instrument of analysis and therefore depended upon the prior establishment of the apparatus as such an instrument. As we have seen, this was a social and rhetorical accomplishment, attained by Davy through building and exploiting a certain relationship with his audience, and through the deployment of potent argumentative techniques. He had had to beat off a series of challenges to his interpretation of the basic experimental phenomenon, the electrolysis of water. Without general agreement that this produced only hydrogen and oxygen, the Voltaic pile could not have functioned as an instrument of analysis. If its workings had remained problematic, the pile could not have been used as a tool of research, at least not with generally recognized results. What was necessary was for the apparatus to be 'black-boxed', in order that the results of its use should achieve acceptance. Davy in fact succeeded in his effort of persuasion, and went on to fulfil the promise that his construction of the instrument had raised. He demonstrated (as Faraday was also to do later), that 'he who proves, discovers' (Gooding, 1985b).

In fact, so completely did Davy's audience accept the instrumental efficacy of the pile that his discoveries with it were widely said to have been expected, given the promise of the device that he now had at his command. In the view of some commentators, mere possession of the instrument guaranteed discoveries. Brougham put the point with characteristic vigour (1808: 395): Davy owed much to 'the powerful instrument which former discoveries had put into his hands'and 'Any man possessed of his habits of labour, and the excellent apparatus of the Royal Institution, could have almost ensured himself a plentiful harvest of discovery'. Although indicative of the degree of trust that Davy's audience placed in his apparatus, this assertion implied that he deserved relatively little credit, compared with Volta, the inventor of the instrument he had used. Exception to this view was taken by John Davy. In a footnote in his edition of his brother's works, John insisted that new applications of existing instruments deserved as much 'honour' as entirely new discoveries (1839-40, 5: 62-3). The insistence is interesting because of what it implies about the status of Davy's instruments, particularly the Voltaic pile: so successfully had he convinced his audience of the power of his instrument that it now seemed to yield discoveries entirely by itself.

This was not a state of affairs with which Davy himself was happy. He regarded discoveries as individual achievements, the property and honour of the discoverer, and a strong part of his claim to immortality (Levere, 1980). In one of his early notebooks he had written: 'A scientific discovery can never be lost; its benefit extends to future ages and the parent of it becomes immortal in his works' (Davy MSS, box 22b, 168). Yet, precisely because Davy had so convincingly argued for the power of his instrument, the discoveries that it yielded became not so much his property, as properties of Nature.

Paradoxically, Brougham's comment also drew attention to the unprecedented concentration of resources that Davy had deployed. His remark implies that Davy's discoveries might only be producible by the uniquely powerful battery of the R.I. and thus that his results could not be replicated by other investigators. This kind of charge worried Davy.

He and his brother repeatedly assured readers that the discoveries had all been achieved using a battery of quite modest size, and that they could very readily be replicated.[21] But, while the vast majority of his audience accepted his discoveries as facts of nature, some objected to the concentration of experimental resources and the curtailment of the facility of replication that had brought them about.

Underlying these tensions were political differences about the proper form of public science. Brougham spoke as a Scottish Whig, who distrusted the metropolitan scientific establishment and its institutions. His own judgment (1810-11: 403) was that Davy had not succumbed to the corrupting atmosphere of the R.I. and had maintained his 'manly' independence. Other observers were not so sure. Davy had abandoned his roots among provincial enlightened intellectuals and had moved into, and exploited to the full, the central institutions of the scientific establishment. Here he formed around himself a new public audience for science and in this context consolidated the rhetorical resources that enabled him to establish the power of his instruments. Those who were politically antipathetic to the fashionable world in which Davy now moved tended to point to the concentration of resources and power that he now had in his hands. For them, Davy's work had made the Voltaic pile less democratically available and more like a private means of discovery. Davy may have convinced many of his British critics of the analytical efficacy of the pile, but (as we shall see in the next section) he was to continue to face objections to other aspects of his work from disgruntled provincial chemists.

Leading French chemists responded to the 1807 Bakerian Lecture with enthusiasm and set about constructing an apparatus with which to replicate, and if possible to challenge, his investigations. The 1806 lecture had been hailed in France and its author awarded a Napoleonic medal for his endorsement of the metallic-contact theory of the pile's action, which was prevalent among French physicists. In 1807 an order was given to construct a new Voltaic pile, of unprecedented power, to compete with that under Davy's command (Gay-Lussac & Thenard, 1811, **1**: x-xi, 1-7; Sutton, 1981). Davy's response was to initiate a subscription fund among the patrons of the R.I. for a new and more powerful battery there. In his petition, presented to the Managers on 11 July 1808, he stressed that national priority in this area of science was in danger of being lost, an outcome that would be 'dishonourable to a nation so great, so powerful, and so rich' (quoted in Jones, 1871: 356). The Proprietors and other subscribers responded by collecting more than one thousand pounds. A pile of two thousand double plates, each of eight inches square, was rapidly built (Davy, 1839-40, **4**: 110-1; **5**: 282-3; Jones, 1871: 355-7). Electrochemistry entered a period in which Davy's supremacy was subjected to a new challenge: international competition.

3. PUBLIC DEMONSTRATIONS AND CHLORINE

CAROLINE: But what was Sir H. Davy's reason for adopting an opinion so contrary to that which had hitherto prevailed?
MRS. B.: There are many circumstances which are favourable to the new doctrine; but the clearest and simplest fact in its support is, that if hydrogen gas and oxy-muriatic gas be mixed together, both these gases disappear, and muriatic acid gas is formed.
EMILY: That seems to be a complete proof; is it not considered as perfectly conclusive?
MRS B.: Not so decisive as it appears at first sight; because it is argued by those who still incline to the old doctrine, that muriatic acid gas, however dry it may be, always contains a certain quantity of water, which is supposed essential to its formation. So that, in the experiment just mentioned, this water is supplied by the union of the hydrogen gas with the oxygen of the oxy-muriatic acid; and therefore the mixture resolves itself into the base of muriatic acid and water, that is, muriatic acid (Marcet, 1817: 148).

When the R.I.'s new battery was finished in 1809, Davy gave a special lecture to introduce it to his audience and to thank the contributors who had made it possible. He described the apparatus as a fitting response to the threat that French experimenters would take the lead in electrochemistry. Individual munificence, spurred by emulation, could match anything the Napoleonic regime could do. He expressed his faith that 'our sciences will always flow from the voluntary efforts of individuals, from whom the support will be an honour--to whom it will be honourable' (1839-40, **8**: 355-6).

In these terms, Davy connected the construction of the new pile with the traditions of English Enlightenment science: a voluntary activity that was dependent on individual patronage, stimulated by emulation and the quest for 'honour'. But, he acknowledged that a change had taken place: his science now drew its strength from a single powerful experimental apparatus which required the widest public support. Widespread curiosity about science would have to be fostered if material support were to continue. Moreover, some coherent direction of the diverse forms of local scientific culture was required since 'To divide and to separate the sources of scientific interest, is to destroy all their just effect'(1839-40, **8**: 357). Public interest had to be focused upon a central locus of scientific activity which could then exploit the material support the public would deliver.

In the following years, Davy worked to mobilize British public opinion in his support, as he found himself embroiled in sustained rivalry with the French chemists Joseph Louis Gay-Lussac and his collaborator Louis-Jacques Thenard. As John Davy noted, Gay-Lussac and Thenard were like 'contending generals', manoeuvring their weapon, the Voltaic battery in Paris, to attack Davy, just as the real generals of England and France were doing in the continuing war between the two nations.[22]

The techniques which Davy used to defend himself in these prolonged scientific battles were simultaneously experimental, rhetorical and social. He tried to win his arguments by doing more experiments than his opponents, exploring more possible claims and counter-claims; he replicated his opponents' work assiduously to demonstrate their 'errors'; and he attempted to sustain a higher degree of precision in all aspects of measurement. The 'facts' produced by experiment would have been impotent without the persuasive force they were given by the use of these techniques. In addition, Davy manipulated chemical language, making several suggestions for changes in nomenclature that would embody his empirical claims. Finally, he buttressed his assertions as to the results of specific experiments by mobilizing various kinds of audiences as witnesses. Davy took advantage of his relations with the general public and specialist chemists by invoking their authority as witnesses to his experiments. He succeeded in turning 'the lever of experiment', by placing its fulcrum on the base of the relationship with his audience that he had so successfully constructed.

To do this, he literally brought the lecture-theatre into the laboratory, assembling at the site of his experiments an audience of expert chemists and prominent patrons. In 1804 the basement laboratory of the R.I. had been enlarged by knocking down a wall dividing it from an adjacent cellar. In the new space a bank of seats was built, capable of accommodating up to 150 spectators, each of whom would have a clear view of the laboratory bench. The declared intention was to create 'a theatre for those who attend the experiments of research' (Jones, 1871: 258).[23] The audience invited there was more select than that which attended in the public lecture theatre upstairs. Members of the R.I.'s Committee of Chemistry were apparently among those who were invited to sessions in the basement laboratory. By invoking this audience as witnesses in his published papers, sometimes in general terms, sometimes as named individuals, Davy brought their authority to bear in support of his experimental claims. For example, the chemist W. H. Pepys and 'a numerous assembly' were invoked to contradict a French claim that potassium was capable of forming a compound with hydrogen

(1839-40, 5: 158n). In response to another reported observation by Gay-Lussac and Thenard, Davy recorded (5: 207):

> I have repeated the experiment...more than twenty times, and often in the presence of some of the most distinguished chemists in this country, from whose acuteness of observation, I hoped no source of error could escape.

Citing respected witnesses to his experiments was one way in which Davy mobilized his audience in specialist controversy. Using this and other techniques which were simultaneously experimental, rhetorical and social, he defended his discovery of potassium and its identification as a metallic element. While the discovery had been widely accepted, challenges to it continued for several years. The Manchester chemist John Dalton was taken to task by the Davys when the second part of his *New System* (1810) identified potassium as a compound (1808-27, 1: 260-3: 2: 53-4; Davy, 1839-40, 5: 322-3; J. Davy, 1811a: 205-6). In 1811, the Edinburgh chemical lecturer John Murray (d. 1820), writing in *Nicholson's*, took up a claim by Claude-Louis Berthollet that potash contained combined water (Murray, 1811b). If this were the case, then its 'decomposition' by the Voltaic pile would not be as Davy had described: the oxygen he had observed could have come from decomposed water and 'potassium' could be a compound of hydrogen with some unknown basis. Responding to Murray, John Davy curtly cited his brother's determination of the water-content of potash, and referred his opponent also to a brief statement by Gay-Lussac and Thenard, in which (he claimed) they accepted the elementary status of potassium (J. Davy, 1811b). Gay-Lussac and Thenard themselves, however, in their *Recherches* (1811), by no means took Davy's side. They insisted that potash and soda did contain water, which would be decomposed by the operation of the pile. It was thus possible that potassium and sodium should continue to be regarded as compounds of the unknown bases of potash and soda with hydrogen (1811, 2: 215-58).

In mounting his defence of potassium and sodium, Davy was pushed into a dilemma as he tried to reconcile his discoveries of two new elements with an explicit methodological commitment to reduce the number of principles in nature. In the 1808 paper in which he criticized the French chemists' view that potassium was a compound of potash with hydrogen, he was prepared to speculate that (like all metals) it might be formed by a union of hydrogen with some unknown basis (1839-40, 5: 131-2). As the assault on his discoveries continued and as the analytical efficacy of the pile was brought into question again, Davy dropped such speculation and insisted upon a rigorous definition of an 'element' as the actually attained limit of chemical analysis.[24]

This definition, which was defended co-extensively with Davy's actual discoveries, linked them with his claim to have a special authority over the instruments of analysis. If one accepted that Davy had a particular knowledge and command of the most appropriate and powerful analytical instruments, one would be inclined to accept his assertions as to what substances were elements. But conversely, if one doubted the efficacy of his instruments one would be likely to dispute some of his identifications of elements and might also conclude that such a definition of an element was unhelpful. Despite its overt epistemological modesty, the definition had to be enforced quite firmly on those who challenged Davy's view of where the limits of chemical analysis lay. This was particularly the case in the controversy which followed his announcement in 1810 of the elementary status of chlorine.

Of all the elements Davy 'discovered', chlorine was the one on which his claim was most open to challenge. The opinion that what had been known as 'oxymuriatic acid' might not be an oxide of muriatic acid, but a simple substance, was first advanced tentatively by Gay-Lussac and Thenard (1809). To them, this was a hypothetical alternative to the traditional view, equally consistent with experimental facts but on balance less simple to reconcile with the remainder of chemical theory. Davy converted French 'opinion' into English 'fact' and in two papers published in July and November 1810 (1839-40, 5: 284-311, 312-48), committed himself to the assertion that oxymuriatic acid was an element.

To defend this claim, Davy brought forward no dramatically new experimental evidence. There was no new substance to be exhibited: the basic phenomenal properties of oxymuriatic acid had been familiar since the description of its preparation by Carl Wilhelm Scheele in 1774. All Davy was able to do was to assert that available techniques had failed to decompose this substance, that it should therefore be considered an element, and that chemical theory could be revised and made consistent with this result. Each step of the reasoning was potentially contentious. In taking the first step, Davy exploited the accepted power of his analytical instruments. He confirmed that oxymuriatic acid could not be reduced by heating with charcoal and added that neither could it be decomposed by the most potent battery at his command (1839-40, 5: 292-3). The implication that it should herefore be regarded as an element required acceptance of his definition of an element. He reminded his readers, 'names should express things and not opinions; and till a body is decompounded, it should be considered as simple' (5: 347n). This conclusion was rendered more plausible by a sketch of how chemical theory could be reconstructed to incorporate the new element. It was portrayed as electro-negative and closely analogous to oxygen; it was capable of combining vigorously with metals and combustible substances, and of displacing oxygen from oxides (5: 292, 296-7, 341-5).

These arguments were supported with the same tactics that Davy had deployed in his previous papers. Witnesses to the experiments were cited. He insisted that what he was reporting were 'facts', and not in the least hypothetical (5: 291, 348n). Above all, he gave the entity a new name, 'chlorine', which implied its elementary nature and obliterated the suggestion that it was an oxide (5: 345). The re-naming strategy clearly owed something to Lavoisier, who had shown very well how adherents could be recruited by getting people to use a new language. But, presumably mindful of this precedent, Davy was careful to validate his change of nomenclature by citing the approval of the specialist chemical community. The change of name was advisable, he proposed, 'to assist the progress of discussion, and to diffuse just ideas on the subject'; he had adopted the term 'chlorine' only after 'consulting some of the most eminent chemical philosophers in the country' (5: 345, 347-8).

These tactics did not guarantee acceptance of Davy's claims. Gay-Lussac and Thenard initially remained unconvinced. In their *Recherches* (1811, 2: 155-66, 171-6), they described the notion that oxymuriatic acid was an element as no more than a plausible hypothesis which was not easily reconciled with Lavoisier's theory of acidity. According to this widely-accepted theory, muriatic and oxymuriatic acids had to contain oxygen, the principle of acidity (Crosland, 1973; Le Grand, 1972). The Swedish chemist J.J. Berzelius also resisted Davy's assertion for a further ten years (Melhado, 1981: 216-7). Brougham, assessing Davy's papers on oxymuriatic acid, judged that he had not entirely proved his point. He lamented that, 'the experiments are not detailed with such minuteness as is essentially necessary, when any new point of doctrine is to be established'; and he resented the resort to new nomenclature, complaining, 'We wish Mr. Davy would attempt to confine himself within the common limits of the language' (1810-11: 406; 1811: 472).

The most sustained challenge to Davy's assertion was made by John Murray, an Edinburgh lecturer and author of two well-known textbooks (Murray, 1801; 1806).[25] In a series of articles published in *Nicholson's* between February 1811 and April 1813, Murray mounted a consistent and plausible attack on Davy's view. John Davy sustained a tenacious defence on his brother's behalf. The controversy reveals much about the rhetoric of argumentation and about the difficulties of securing agreement on interpretation of experiments. It also reveals the uses which could be made of a public audience in attempting to resolve specialist disputes.

Murray's challenge led off from the supposition that acids, such as muriatic acid, required combined water in order to exist as independent entities. On this basis, the experiments on which Davy rested his theory could be given an alternative explanation. For example, oxymuriatic acid combined with hydrogen to form muriatic acid because its oxygen formed the water which was part of the final product. Charcoal would not be expected to reduce oxymuriatic acid to muriatic acid because charcoal could not provide the water that was needed to constitute muriatic acid in its gaseous form (Murray 1811a). John Davy in reply tied the defence of chlorine to that of his brother's definition of an element, a definition he described as 'a fundamental principle of modern chemistry'. He denounced Murray's muriatic acid with combined water, as 'an imaginary body, that I confess I am altogether ignorant of' (1811a: 193, 199).

One of the experiments discussed was that in which Murray claimed to have used oxymuriatic acid to provide the oxygen to convert 'carbonic oxide' [CO] to 'carbonic acid' [CO_2]. In Davy's account, in which oxymuriatic acid was actually the element chlorine, this was impossible. Not surprisingly, when the Davys repeated Murray's experiment, with what they considered appropriate precautions to purify and dry the gases, they found no carbonic acid was formed. In John Davy's view (1811a: 201-2), Murray's gases must have been contaminated with water and/or atmospheric oxygen. Replication of Murray's experiments, with more rigorous procedures, had supposedly produced results consistent with Davy's new theory.

The Davys' strategy, to 'replicate' their opponent's experiments in order to expose his 'errors', was one Murray found difficult to handle. He could only respond by repeating his own experiments, with 'every attention...to ensure accuracy', and re-iterating that the original results stood (Murray, 1811c). Murray's argumentative style relied on the denial, not of 'facts', but of the Davys' 'inferences' from them. He sought to show that the traditional account of oxymuriatic acid was preferable because of its coherence with established chemical theory. Hence, he was committed to portraying both accounts as equally 'hypothetical' constructs. Murray (1811d: 195) insisted:

...the very possibility of the proposition being called in question without any doubt being expressed of the accuracy of the experiment on which it rests is a sufficient proof, that it is not a simple expression of the fact, as he [John Davy] and his brother suppose, but an inference from the fact.

John Davy drew a different line between 'facts' (which included deductions from experiments) and 'hypotheses' (J. Davy, 1811a: 194, 205-6; 1811c: 39-41). He defended his brother's discovery of chlorine as an experimental 'fact' and hence was obliged to deny the factuality of many of Murray's experimental claims, displaying them as mistaken inferences from improperly conducted experiments. For example, on Murray's assertion that muriatic acid when combined with potash released water, J. Davy commented (1811c: 40): 'if the above is a simple expression of facts, the theory which expresses those facts must be correct. But I have not been able to witness such facts'.

The early stages of the debate thus proceeded in a manner which both protagonists realized was likely to be inconclusive. While John Davy complained that Murray was denying plain experimental 'facts', Murray continued to uphold his own results and the validity of his inferences from them. He complained of Davy's dogmatic style of expression, while presenting himself as open-minded and tolerant of alternative theories. The Davys mobilized the preponderance of prestige witnesses (including such members of the Chemical Committee as Charles Hatchett and William Thomas Brande), as against Murray's assistant Mr. Ellis and 'some other friends' (J. Davy, 1811c: 42; Murray, 1811d: 190); but the debate could not be resolved by these means.

The stalemate was only broken when John Davy recruited powerful new 'allies' to support his case. In September 1811 he declared triumphantly (1811d: 30):

I have now to announce the existence of a new acid gas, which operated in Mr. Murray's experiment, without his knowledge of its existence, and was the cause of those phenomena which he erroneously attributed to the formation of carbonic acid gas.

John Davy's act of producing a new gas (subsequently dubbed 'phosgene') and asserting its (unknown) presence in his opponent's experiments proved a powerful move in the dispute. He had already drawn Murray's attention to the possible complications caused by a new compound of oxymuriatic acid and oxygen, recently described by Humphry Davy in a paper read to the Royal Society on 21 February 1811 (1839-40, 5: 349-57). Murray had denied (1811d: 200-1) that that gas, subsequently named 'euchlorine', had been present in his experiments. Now he was faced with another new gas, 'a new and peculiar compound of carbonic oxide and oximuriatic gas', which behaved in a way liable to be confused with the 'carbonic acid' that Murray claimed a mixture of those two gases produced (J. Davy, 1811d: 30). Murray's response was tentative and defensive. He recorded that he could not produce the new gas in the way John Davy had described but, pending the publication of a fuller description of its preparation, he had to accept that there might be 'some peculiarity necessary to the success of Mr. J. Davy's experiment'. He went on, 'I know sufficiently the disadvantage to which any experimentalist is subjected, who undertakes the examination of experiments of which only a general account is given' (1811e: 227).

Murray retreated to consideration of another experiment which seemed to offer the possibility of a decisive resolution of the whole debate: the reaction of muriatic acid gas with ammonia. According to Davy's theory, neither of these gases contained water, so nor could the salt which was the product of their combination. Murray claimed, however, that muriatic acid did contain water and that this could be established by extracting water from the product salt. He wrote, 'This experiment then has the advantage of being conclusive on the subject of the present discussion; the state of the fact only requires to be ascertained, and with due precaution this is not difficult of attainment' (1812a: 131). Murray spoke too soon. Concentrating attention on the reaction of muriatic acid with ammonia focused the debate on what promised to be 'an *experimentum crucis*' but did not in the event resolve it. John Davy rose to the challenge and announced that the water which Murray had found in the product ('muriate of ammonia') had been absorbed from the atmosphere or from insufficiently dried reactant gases: the salt, he claimed, was deliquescent (J. Davy, 1812). Murray in reply denied this, pointing out that the salt neither became damp nor showed any gain in weight on exposure to air (1812b). He also drew support from a successful replication of his experiment, reported from the Liverpool Literary and Philosophical Society, by the physicians John Bostock and Thomas Stewart Traill (Bostock, 1812).

Bostock lent further support to Murray's cause by objecting strongly when the first volume of H. Davy's *Elements* (1812) presented the question of chlorine in an entirely one-sided way. In what was proposed as an authoritative textbook, Davy stated that the traditional view of the composition of the gas was 'now universally given up' (1839-40, **4**: 177). He ascribed the intransigence of certain (unnamed) Scots and Frenchmen to their belief in a fictitious entity called 'dry muriatic acid'. Bostock expressed resentment of the dogmatic tone Davy had adopted in defence of his 'fanciful' beliefs and particularly criticized the construction of a nomenclature based on the chlorine theory (1813: 152):

> ...to employ such [terms] as are at present the subject of controversy, and even to make them the basis of a nomenclature, can only tend to impede the progress of science, and lead us to suppose that we have acquired knowledge when in fact we have only learnt a new language.[26]

Apart from producing a textbook, the Davys were manoeuvring in other ways to bring the controversy to a favourable resolution. In January 1813 John published 'An Account of an Experiment made in the College Laboratory, Edinburgh', describing what was clearly meant to be a decisive refutation of Murray's arguments: a version of the 'crucial experiment' performed by Humphry in Murray's own city, with an assembly of illustrious witnesses including Thomas Charles Hope (Professor of Chemistry at Edinburgh University), Sir George Mackenzie, John Playfair and 'some other gentlemen'. Ammonia and muriatic acid gases were dried and then allowed to mix. The product salt was heated; a tiny amount of dew was given off. This was judged far too small a quantity of water to support Murray's case and was ascribed to 'uncombined moisture derived from various sources', such as the mercury over which the gases were mixed. John (1813) described the experiment curtly and factually and with a lofty air of finality refused to reply to any further objections by Murray.

Murray was not to be beaten in this manner. A few months later, he hit back with his objections to the Davys' experiment and (having taken a leaf from his opponents' book) a description of his own 'replication' of it. The Davys had employed an inappropriate mode of heating the salt, he complained, which diminished the chances of extracting what would only be expected to be a small quantity of water. When he had repeated the experiment, with some of the same witnesses present, 'The quantity of water, Dr. Hope was satisfied, appeared considerably larger than in Sir Humphry's experiment'. Calculations about the expected quantity were beside the point, Murray suggested, since the reaction of muriatic acid with ammonia was not the one best suited to reveal how much combined water the acid contained. The essential point was that water *was* found: 'The production of any water is incompatible with Sir Humphry's hypothesis, and therefore refutes it' (1813: 266, 270). True to his promise, John Davy declined to reply.

The controversy between Murray and the Davys invites analysis in terms derived from the sociology of modern science. The inconclusive nature of the experiments that were performed illustrates what H.M.

Collins (1985) has called 'the experimenters' regress'. Rather than being astonished, as contemporaries apparently were, that the available experimental 'facts' seemed equally compatible with beliefs on both sides of the debate, we should note Collins's observation that a sufficiently determined and resourceful opponent can always find ways to challenge any empirical claim. Experiments themselves cannot convince someone who refuses to assent to them and such a person can always stipulate grounds for their refusal. Collins also argues that it is impossible to specify *a priori* what constitutes a replication. Thus, there always remain grounds for disagreement about whether a 'confirmation' or a 'falsification' has been achieved or whether or not the experiment has been properly replicated. 'Successful' or 'failed' replications, 'confirmations' or 'falsifications', have to be decided socially *via* processes of negotiation and persuasion. Since the author of a written account of an experiment cannot control the actions of another experimenter, both actors retain space in which to argue that their experiments differ in some important respect. Thus, in a dispute in which experiments are mobilized to support alternative theoretical accounts, they cannot *per se* be decisive. Although each actor may designate his opponent's experiments either as successful 'replications' or as 'failures', the opponent can always refuse assent to this and insist that they actually confirm his own theory.

Thus, although we can see that the Davys' strategy of supposedly 'replicating' Murray's experiments in order to show where he went wrong might in other circumstances have been effective, we can also see why in this instance it failed to produce a satisfactory resolution of the controversy. To call Murray's bluff by accepting his suggestion for a 'crucial experiment' and performing it publicly in his own city was a bold move by the Davys and one which in their opinion was crowned with success. But, Murray retained room for manoeuvre and exploited it resourcefully: he denied that the experiment had been properly conducted and repeated it with an outcome more favourable to his point of view.

There was no agreement between the parties that the dispute had been concluded. Later accounts by the Davys presented as accepted fact that Murray had been misled by the unknown presence of euchlorine and phosgene in his early experiments and that the 'crucial experiment' with muriatic acid and ammonia had decisively demonstrated his error (1839-40, 1: 123; 5: 347-8n). Murray never accepted that the dispute had ended in this way. He continued to reject the discovery of the two new gases as irrelevant to his experiments and to insist that the claimed outcome of the 'crucial experiment' had subsequently been reversed. In the third edition of his *Elements* (1814, 1: 484-500), he still gave an even-handed description of the two 'hypotheses' regarding oxymuriatic acid, refusing to accept that the Davys had proved their case. He maintained this position through to the fourth edition of his *System* (1819, 2: 416-78), published the year before his death. He had also reported (1818) further experiments to show that muriatic acid gas contained combined water, thus re-iterating the evidence for judging the chlorine theory fallacious. In Murray's view, the controversy had not been resolved.

Who *was* convinced about the chlorine theory and how were they persuaded? The Davys repeatedly performed public experiments to convince audiences of the truth of their claims but the degree to which these were successful in persuading specialist chemists is doubtful. Humphry Davy's command of his audiences was a resource to which Murray had no counterpart. Hence the brothers' enrolment of expert witnesses, a contest in which (as we have seen) they gained the advantage. Appeal was also made to a wider public. In February 1812 Humphry showed an audience at the R.I. what was wrong with Murray's experiment with muriatic acid and ammonia, a display which was described in a number of monthly journals:

> ...the Professor clearly proved, that the presence of water was owing to the hygrometric qualities of the salt, which, when exposed to the atmosphere for an instant, absorbs moisture directly; and he showed an experiment, in which, when muriatic gas and ammonia were combined out of the atmosphere and heated, not an atom of water could be procured from them.[27]

In the terms of a distinction that Collins (1988) has recently made in a discussion of public science, this exhibition would have to be classified as a 'demonstration' rather than an 'experiment'. It is hard to see how a sceptical and informed onlooker could have been *shown* that 'not an atom of water' was produced. Murray, had he been present, would presumably not have been convinced by such a display. Less well informed and less committed observers could however be expected to be persuaded and it was presumably with this aim that Davy presented such demonstrations. Throughout the controversy, he described his view of chlorine in his R.I. lectures as an established fact supported with conclusive experimental manipulations (*e.g.*, 1839-40, 8: 311-9). Skating over the areas of dispute, he showed his public audiences demonstrations that supposedly gave unambiguous support to his doctrines. He exploited their chauvinism also, presenting his theory of chlorine as an alternative to a pernicious and erroneous French system of chemistry (*e.g.*, **8**: 318).

Specialists could find grounds for dismissing such demonstrations as inconclusive. Murray himself, who found ways of criticizing the Davys' designated 'crucial experiment' in Edinburgh, was certainly not persuaded by anything done in the theatre of the R.I. Thomas Thomson, who was much less committed than Murray to opposing the chlorine theory, likewise hesitated (1814: 14) to judge the 'intricate' matter of the outcome of the Edinburgh demonstration. T.C. Hope, a specialist witness at the Davys' 'crucial experiment', later endorsed the alternative outcome of Murray's 'replication'. Those with significant knowledge of the issues at stake were clearly not as easily persuaded by experimental demonstrations as were members of the general public.

It seems that most specialist chemists remained unconvinced by the chlorine theory until well after the dispute between the Davys and Murray had been abandoned. Acceptance does not seem to have been widespread among chemists until about 1816-18 (Le Grand, 1974b: 224-5). Three factors can be identified which swung support behind Davy's view at that time. First was the discovery of iodine, announced triumphantly by Gay-Lussac and Thenard in August 1814. As the properties of this surprising new substance were explored, it began to appear that it was an element which shared many of the attributes of chlorine.[28] The plausibility of the view that chlorine was itself an element was thus increased. Second, following from this, was the apparent shift of position by the leading French chemists, who gave their allegiance to the chlorine theory and even pressed their claim to have originated it (Thomson, 1816: 28). Third, Davy himself continued to build support for his view. While overtly dismissive of challenges to the chlorine theory, Davy tacitly acknowledged that the debate was still open by publishing a series of papers on iodine and fluorine compounds (1814a; 1814b; 1814c; 1815) which helped strengthen perceived analogies between the newly identified elements and chlorine.

The movement of opinion in the specialist community in favour of Davy's view was chronicled by Thomson in the annual reviews of developments in science published in his journal. In 1814, Thomson was still expressing serious reservations about Davy's theory. The following year he reported the recent work on iodine and on fluorine compounds but it was only in 1816 that he related these developments to the chlorine theory, which he now described as 'pretty generally admitted' (1816). Accordingly, in 1817 the fifth edition of Thomson's *System* was largely re-written to incorporate what he described as the revolutionary new doctrine (1817, 1: vii, 184-5). Other authors were more cautious. Marcet (1817) was still recommending readers to wait for a more decisive judgment by specialist chemists. In Marcet's dialogues, the pupils 'Caroline' and 'Emily' were told by their mentor 'Mrs. B.' that, 'The new doctrine has certainly gained ground very rapidly, and may be considered as nearly established; but several competent judges still refuse their assent to it' (2: 148-50). Samuel Parkes's *Chemical Catechism* was revised (1818: 167, 485-8) to incorporate Davy's theory and to include a discussion of the properties of iodine. The same year, William Henry endorsed the chlorine theory, noting that it had been rendered much more plausible by the discovery of the apparently analogous new element (1818, 1: v, 419-21, 425).

By 1818 Davy had good reason to be confident of the support of his audiences, both specialist and public, for his theory. In a paper read to the Royal Society, he was able magisterially to dismiss experiments reported by John Murray and Andrew Ure to the Royal Society of Edinburgh (Murray, 1818: 291-301, 316). They claimed to have shown experimentally that water could be produced by passing muriatic acid gas over heated iron, implying that chlorine had to be a compound containing oxygen. Davy (1839-40, 5: 524) was dismissive of these claims: 'To take up the time of the society by long experimental details and theoretical speculations on such an occasion, will be unnecessary; I shall therefore only transiently mention the sources of error....' The sources Davy mentioned had a familiar ring: he asserted that the oxygen to form the water was derived from lead oxide or alkali in the flint glass of the apparatus, or from the atmospheric air remaining in the reaction tube.

Davy's credit with his audience was apparently high enough at this point for his assurances to be accepted without much question. William Henry (1818, 1: 425) and Thomas Thomson (1818a: 220; 1818b: 450) both thought his response was adequate to put the matter beyond doubt. A reviewer in the *Monthly Review* (1818: 189), while noting that the controversy regarding the nature of chlorine was not yet closed, also found Davy's reply to the Murray-Ure experiments convincing. Clearly the climate of informed opinion was much more favourably disposed towards his view of chlorine as an element than it had been only three or four years previously.

4. CONCLUSION

These two cases from Davy's career suggest some answers to the question with which I began: How can experiments be made convincing?

They indicate some of the potential and also some of the limitations of attempts to use public experimental demonstrations to establish claims to scientific knowledge, to turn the 'lever of experiment' on the fulcrum of a certain relationship with a public audience.

Davy's tactics in the debate with Murray parallel those he had successfully employed earlier against Gay-Lussac and Thenard and against the other early investigators of the Voltaic pile. In those, he had used forceful rhetorical techniques, forged in the context of appeal to a public audience, and had also invoked the authority of that audience itself to support his descriptions of experiments. By these methods his experiments had gained meaning as supports for his theoretical claims: the Voltaic pile was established as a powerful analytical instrument and the elements sodium and potassium were constituted and defended as important discoveries. In the case of chlorine such tactics were not successful, at least not until well after the dispute with Murray. The Davys' experiments, even those showing purportedly conclusive results before public audiences, were not accepted by their opponent.

One can understand in terms of Collins's analysis (1988) how Murray was able to retain freedom to argue his position. Public 'demonstrations' may be clearly distinguished from true 'experiments', and are likely to be regarded by informed observers as much less convincing. Specialists are only likely to be persuaded by procedures of dialogue within their community, in which all qualified participants have equal ability to perform experiments. Investigators who are perceived as retaining privileged access to resources, or who make what seem to be illegitimate appeals to the judgment of public audiences, may be distrusted by their peers.

There were reasons for such distrust of Davy, arising out of the social and political context in which he forged his career. He had abandoned his roots in the English provincial Enlightenment to move to the fashionable culture of the capital and the heart of the metropolitan scientific establishment. He could be seen as having turned his back on his scientific mentors, Beddoes and Priestley, on their progressive 'Country' politics, and on the democratic form of public culture of science that they espoused. Brougham's comments on Davy's scientific achievements seem to voice the reservations to which such suspicions would give rise. Brougham, an enlightened Scottish Whig, initially expressed apprehension that Davy might have been corrupted by the fashionable atmosphere of the R.I. He was later willing to accept the discoveries of the two new metallic elements but drew attention to the concentration of instrumental resources that had brought them about. When Davy launched his chlorine theory, Brougham protested at his dogmatic tone and at the perversion of a common chemical vocabulary into the service of an unjustified supposition. Similar reservations expressed by Bostock, Henry and Thomson may also have stemmed from distrust felt among liberal provincial intellectuals of Davy's moves to entrench himself in the metropolitan establishment.

On the other hand, Davy's considerable scientific achievements cannot be neglected, and the role in them of the social context in which he worked should not be ignored. The centralization of resources through focusing public support on the R.I. yielded remarkable dividends. Davy's Voltaic battery was consolidated and augmented, both in material construction and in degree of public acceptance of its results. Public experimental demonstrations had an important role in this process. And, in international competition with the French chemists, Davy's strategy brought victories comparable with those being gained by English generals on the field of battle. If Davy were a general of science, planning grand strategy, mobilizing the heavy artillery, consolidating the lines of supply, and pressing on regardless of mutterings in the ranks, then that was apparently what was required.

Looking at Davy in this way lends support to, and helps to extend, Ludwik Fleck's analysis of 'the general epistemological significance of popular science'. Fleck proposed (1979: 106, 113-5) that 'certainty, simplicity, vividness originate in popular knowledge'; that they are generated by processes of 'exoteric' communication with a public audience (*cf.* Whitley, 1985). We have seen Davy enhancing the 'vividness' of his experiments, in order to present them in his lectures, simplifying complex and controversial subjects, and thereby consolidating the 'certainty' of disputed 'facts'. This simplified and more certain knowledge can then be fed back into 'esoteric' discourse among specialists. The resolution of specialist disputes, and the firming-up of scientific knowledge thereby, benefits from the continual circulation of discourse from the 'esoteric' to the 'exoteric' realms, and back again. Davy's achievement instantiates Fleck's claim; it shows how appeal to a public audience can consolidate and strengthen knowledge in the expert realm.

The case of Davy also demonstrates how the importance of this process might have been accentuated at certain historical junctures. Faced with the failure of some Enlightenment forms of public science, Davy mobilized a new audience in support of a newly reconstructed and more resource-intensive discipline of chemistry. His concentration of instrumental power and his presumption of public support for his experimental claims were frequently resisted, particularly by those who recalled the more egalitarian and democratic ideals of public science that Priestley had upheld. The Enlightenment, however, was over. Davy had anticipated that nineteenth-century chemistry was to depend upon substantially increased resources of apparatus and skills and he saw how the public audience for chemistry could be extended to deliver the extra support chemists would now require. That support took the form of financial contributions and also of public 'credit' for the authority of the scientist and the validity of his claims to knowledge. Davy had shown how a public audience could be a weapon to be wielded in the battles through which scientific knowledge was established.

NOTES

The contents of this paper appear in modified form in my forthcoming book, *Science as public culture: Chemistry and Enlightenment in Britain, 1760-1820* (© Cambridge University Press, reprinted with the permission of Cambridge University Press).

1. Representative sociological studies include: Collins, 1985; Latour, 1987; Latour & Woolgar, 1986; Pinch, 1986. Among historical studies informed by this work see, in particular: Gooding, 1985a; Shapin, 1979; 1984; Shapin & Schaffer, 1985.
2. This account is indebted to previous historical research on Davy. A traditional appraisal of his achievement is given in Partington, 1961-70, **4**: 29-76. The most comprehensive study of his chemical theory is Ziemacki, 1974. Other useful studies of his natural philosophy include Levere, 1971, ch. 2; Knight, 1978, chs. 2 & 3; and the studies in Forgan, 1980. Biographies include Hartley, 1966; Treneer, 1963. I have also found the numerous articles of June Fullmer particularly informative on Davy's life and career; see Fullmer, 1962; 1964; 1980.
3. Thomas Beddoes, writing in the *Monthly review* (1801), quoted in Stansfield, 1984, 110. On the 'end of the Enlightenment' see Darnton, 1970.
4. Davy's speculations on galvanism are found, *e.g.*, in MSS of Sir Humphry Davy, Royal Institution, London (henceforth cited as Davy MSS), box 20b, p. 108, and box 22b, pp. 36-7, 137-8, 155-6. Quoted by courtesy of the Director of the Royal Institution.
5. Southey to Davy, 26 July 1800, in Davy MSS, box 9, p. 109.
6. Volta described the construction of his pile in a letter to Sir Joseph Banks, read to the Royal Society on 26 June 1800, and subsequently published as Volta, 1800. Nicholson and Carlisle's experiment was described in Nicholson, 1800-01, published in July 1800. The speed with which Volta's apparatus was replicated and applied probably owes much to the previous attempts to decompose water using frictional electricity, such as Pearson, 1797-8. The term 'electrode' which I have used is anachronistic. Contemporary investigators tended to refer to 'wires', 'points' or 'poles'.
7. On the 'black-boxing' of instruments, whereby they become accepted as devices for producing new knowledge, the workings of which are not brought into question, see: Pinch, 1986, esp. 211-4; Collins, 1985; Latour, 1987, esp. 2-3, 80-3, 131.
8. Nicholson (1800-01) gave easy-to-follow instructions for reproducing his apparatus, recommending the use of silver half-crowns and discs of zinc, and giving an address of

a supplier from whom the latter could be obtained. For examples of other experimenters who followed his recommendations, see: Henry, 1800-01; Haldane, 1800-01.
9. The extent of reproduction of the pile is suggested by a description of research at the Chemical Society of Edinburgh, in a letter of John Thomson to John Allen, 12 June 1801: 'Lord Lauderdale and I made the galvanic experiment last week, and I exhibited it to the Society on Saturday. We are getting tubes with gold wires and glass stoppers to try its effects on caustic liquids, and are getting a very broad plate of zinc made, to try whether the increase in power be in proportion to the increase of surface. In that case, his Lordship's whole service of plate will be converted into a galvanic battery'. (Quoted in Thomson, 1832-59, **2**: 17).
10. Other reports concerning the production of acids and alkalis, by the Italians and other investigators, can be found in the *Philosophical magazine*, **9** (1801): 181-5; **24** (1806): 176-80, 185; **25** (1806): 57-66, 130-42, 364-5; **27** (1807): 260-1, 338-9, 339-43.
11. On Sylvester's theory of the oxidizing power of electricity, and for a comparison with the rather different interpretation of his results by Pacchiani, see Le Grand, 1974b: 216-8.
12. Gregory Watt to Davy, 7 February 1801, Davy MSS, box 9, p. 87.
13. The audience at Davy's lectures is discussed in Foote, 1952; Berman, 1978; Yearley, 1985; and in the papers by Fullmer and Siegfried in Forgan, 1980. Press reporting and Davy's attitudes to it are mentioned in Fullmer, 1962: 150-2. Reviews of Davy's published (and some unpublished) lectures are listed in Fullmer's very helpful bibliography (1969).
14. [Marcet], 1817: esp. vii, ix, 3-4. See also the remarks of Joseph Priestley, 1778, 137-8; Adam Walker, 1807: v; and Leonard Horner: 'Your chemists and metaphysicians in petticoats are altogether out of nature -- that is, when they make a trade or distinction of such pursuits -- but when they take a little general learning as an accomplishment they keep it in very tolerable order' (Horner to J.A. Murray, 15 November 1804, quoted in Jones, 1871: 297-8, 264).
15. On Davy's cultivated oratorical style, see Davy, 1839-40, **1**: 91-3; Hartley, 1966: 46; Treneer, 1963: 88-9. For remarks on his appearance, Fullmer, 1962: 158; Treneer, 1963: 86; *Philosophical magazine* **9** (1801): 281-2. For examples of spectacular demonstrations see: Foote, 1952: 11; Levere, 1981: 30; Davy, 1839-40, **2**: 188-209, 211-3.
16. See, for example, Davy MSS, box 13c, p. 32. Davy's notions of 'genius' are discussed in Levere, 1980.
17. The incident on 20 June 1801 is described in a report in the *Philosophical magazine* **10** (1801): 86-7. The 'select party' which met to continue the experiments on 23 June comprised Count Rumford, Sir Charles Blagden, W.H. Wollaston and others. Gillray's cartoon is discussed in Wright & Evans, 1851: 468-9; it has been related to an occasion in March 1800, when Davy assisted Thomas Garnett in demonstrating nitrous oxide at the R.I., in June Fullmer's letter to *Scientific american* **203** (no. 2) (August, 1960): 12-14.
18. Davy, 1839-40, **5**: 2-4. These technical refinements were not presented by Davy as tactical moves in a debate, but they did serve that function. Like other technical innovations, such as increasing the precision of measurement, making experiments harder to perform could be a way of establishing knowledge by reducing the number of possible alternative claims.
19. See the remarks of the reviewer in the *Philosophical magazine* **26** (1806-1807): 181, 266-7, 269; Bostock (1807); Brougham (1807-1808).
20. See reviews in *Philosophical magazine* **29** (1807-1808): 180-3; *Monthly review* **57** (1808): 225-7; *Edinburgh review* **12** (1808): 394-401.

21. See, for example, the footnotes in Davy, 1839-40, 5: 62-3, 106-7; **8**: 355. The assertion that Davy's decomposition of the alkalies had been effected using an especially powerful battery at the R.I. nevertheless continued to be made, *e.g.* in John & George Murray, 1834: 317-320.
22. The long-standing rivalry between Davy and the French chemists is described in Crosland, 1978; 1980. John Davy's military analogy is quoted in Crosland, 1978: 76. The background of the Napoleonic wars is a pertinent one. In his lectures, Davy described his debates with Gay-Lussac and Thenard in terms of a rivalry between English and French schools of chemistry; and, as Fullmer has shown (1962: 150-1), he also had to defuse criticism that he was lacking in patriotism for having accepted the award of the Napoleonic medal.
23. For a brief description, with perspective view and plan, of the modified basement laboratory, see Chilton & Coley, 1980.
24. Davy was here turning to his own advantage the definition of an element which had originally been enunciated by Lavoisier. See Davy, 1839-40, 5: 527n.; J. Davy, 1811a: 193; and the discussion in Le Grand, 1974a: 62.
25. This John Murray should not be confused with his son, John Murray (1798-1873), or his namesake, John Murray (1786?-1851). All three are listed in the *Dictionary of national biography*. The Murray in question here gave private lectures in Edinburgh, which are mentioned in Morrell, 1972: 13. There is an obituary notice in the *Gentleman's magazine* **90** (1820), ii: 185-6.
26. Comparable remarks were made by other reviewers in *Annals of philosophy* **1** (1813): 371-7; *Quarterly review* **8** (1812): 65-86; *Philosophical magazine* **40** (1812): 145-51, 297-307, 434-44.
27. Quoted from *Monthly magazine* **33** (1812): 260. Identical wording was used in the reports in *Philosophical magazine* **39** (1812): 132-40 and *Nicholson's* **31** (1812): 236-7.
28. On the discovery of iodine, see Crosland, 1978: 80-7. Murray (1819, 2: 426) acknowledged that this discovery had persuaded many others of the elementary status of chlorine.

REFERENCES

[Anon.] (1800). *The sceptic*. Retford: E. Peart.

Berman, M. (1978). *Social change and scientific organisation: The Royal Institution, 1799-1844*. Ithaca: Cornell University Press.

Bostock, J. (1807). Review of Davy's Bakerian Lecture. *The monthly review*, **54**, 1-4.

Bostock, J. (1812). Experiment to prove whether water be produced in the combination of muriatic acid gas and ammoniacal gas. *Nicholson's journal* [hereafter, *Nicholson's*], (ser. 2), **32**, 18-21.

[Bostock, J.]. (1813). Review of Davy's *Elements of chemical philosophy. The monthly review*, **72**, 148-58.

Brougham, H. (1807-1808). Review of Davy's 1806 Bakerian Lecture. *The Edinburgh review*, **11**, 390-8.

Brougham, H. (1808). Review of Davy's 1807 Bakerian Lecture. *The Edinburgh review*, **12**, 394-401.

Brougham, H. (1810-11). Review of Davy's paper 'Researches on oxymuriatic acid'. *The Edinburgh review* **17**, 402-9.

Brougham, H. (1811). Review of Davy's 1810 Bakerian Lecture. *The Edinburgh review* **18**, 470-80.

Chilton, D. & Coley, N.G. (1980). The laboratories of the Royal Institution in the nineteenth century. *Ambix*, **27**, 173-203.

Collins, H.M. (1985). *Changing order: Replication and induction in scientific practice.* Beverly Hills: Sage.

Collins, H.M. (1988). Public experiments and displays of virtuosity: The core-set revisited. *Social studies of science*, **18**, 725-48.

Crosland, M. (1973). Lavoisier's theory of acidity. *Isis*, **64**, 306-25.

Crosland, M. (1978). *Gay-Lussac: scientist and bourgeois.* Cambridge: Cambridge University Press.

Crosland, M. (1980). Davy and Gay-Lussac: Competitionand contrast. In Forgan (1980), pp. 95-120.

Cruikshank, W. (1800-01). Some experiments and observations on galvanic electricity. *Nicholson's*, (ser. 1), **4**, 187-91, 254-64.

Cuthbertson, J. (1806). Untitled letter. *Philosophical magazine*, **24**, 170-1.

Dalton, J. (1808-27). *A new system of chemical philosophy.* 2 vols., Manchester and London: Russell, Bickerstaff & Wilson.

Darnton, R. (1970). *Mesmerism and the end of the Enlightenment in France.* New York: Schocken.

Davy, H. (1814a). An account of some new experiments on the fluoric compounds. *Philosophical transactions*, **104**, 62-73.

Davy, H. (1814b). Some experiments and observationson a new substance which becomes a violet coloured gas by heat. *Philosophical transactions*, **104**, 74-93.

Davy, H. (1814c). Further experiments and observations on Iodine. *Philosophical transactions*, **104**, 487-507.

Davy, H. (1815). Some experiments on a solid compound of Iodine and Oxygene, and on its chemical agencies, *Philosophical transactions*, **105**, 203-13.

Davy, H. (1839-40). *The collected works of Sir Humphry Davy, Bart., LL.D., F.R.S.* 9 vols., ed. J. Davy, London: Smith, Elder & Co.

Davy, J. (1811a). Some remarks on the observations and experiments of Mr Murray. *Nicholson's* (ser. 2), **28**, 193-206.

Davy, J. (1811b). An answer to Mr. Murray's observations on the nature of Potassium and Sodium. *Nicholson's* (ser. 2), **29**, 35-9.

Davy, J. (1811c). On the nature of oximuriatic gas. *Nicholson's* (ser. 2), **29**, 39-44.

Davy, J. (1811d). An account of a new gas, with a reply to Mr. Murray's last observations. *Nicholson's* (ser. 2), **30**, 28-33.

Davy, J. (1812). On the nature of oximuriatic acid. *Nicholson's* (ser. 2), **31**, 310-6.

Davy, J. (1813). An account of an experiment made in the college laboratory, Edinburgh. *Nicholson's* (ser. 2), **34**, 68-72.

'Experimentalist, An'. (1800-01). On the chemical effects of the pile of Volta. *Nicholson's*, **4**, 472-3, 514-5.

Fleck, L. (1979). *Genesis and development of a scientific fact.* T.J. Trenn & R.K. Merton, eds., Chicago: University of Chicago Press.

Foote, G.A. (1952). Sir Humphry Davy and his audience at the Royal Institution. *Isis*, **43**, 6-12.

Forgan, S., ed. (1980). *Science and the sons of genius: Studies on Humphry Davy.* London: Science Reviews.

Fullmer, J.Z. (1962). Humphry Davy's adversaries. *Chymia*, **8**, 147-64.

Fullmer, J.Z. (1964). Humphry Davy and the gunpowder manufactory. *Annals of science*, **20**, 165-94.

Fullmer, J.Z. (1969). *Sir Humphry Davy's published works*. Cambridge, MA: Harvard University Press.

Fullmer, J.Z. (1980). Humphry Davy, reformer. In Forgan (1980), 59-94.

'G.M., Dr.' (1800-01). On the state of galvanism and other scientific pursuits in Germany. *Nicholson's* (ser. 2), **4**, 511-3.

Gay-Lussac, J.L. & Thenard, L.J. (1809). De la nature et des proprietes de l'acide muriatique oxigene. *Memoires de physique et de chimie de la Societe d'Arcueil*, **2**, 339-58.

Gay-Lussac, J.L. & Thenard, L.J. (1811). *Recherches physico-chimiques, faites a l'occasion de la grande batterie voltaique donnee par S.M.I. et R. a l'Ecole Polytechnique*. 2 vols., Paris: Deterville.

Gooding, D. (1985a). 'In nature's school': Faraday as an experimentalist. In *Faraday rediscovered*, ed. D. Gooding & F.A.J.L. James, pp. 105-35. London: Macmillan.

Gooding, D. (1985b). 'He who proves, discovers': John Herschel, William Pepys and the Faraday effect. *Notes and records of the Royal Society of London*, **39**, 229-44.

Haldane, H. (1800-01). Experiments and observations made with the newly discovered metallic pile of Sig. Volta. *Nicholson's* (ser. 1), **4**, 241-5.

Hartley, H. (1972). *Humphry Davy*. Wakefield: EP Publishing.

Henry, W. (1800-01). Experiments on the chemical effects of galvanic electricity. *Nicholson's* (ser. 1), **4**, 223-6.

Henry, W. (1818). *The elements of experimental chemistry*. 8th edn., 2 vols., London: Baldwin, Cradock & Joy.

Jones, H.B. (1871). *The Royal Institution: Its founder and first professors*. London: Longmans, Green and Co.

Kipnis, N. (1987). Luigi Galvani and the debate on animal electricity. *Annals of science*, **44**, 107-42.

Knight, D.M. (1978). *The transcendental part of chemistry*. Folkestone: Dawson.

Latour, B. (1987). *Science in action: How to follow scientists and engineers through society*. Milton Keynes: Open University Press.

Latour, B. & Woolgar, S. (1986). *Laboratory life: The construction of scientific facts*. Princeton: Princeton University Press.

Le Grand, H.E. (1972). Lavoisier's oxygen theory of acidity. *Annals of science*, **29**, 1-18.

Le Grand, H.E. (1974a). Determination of the composition of the fixed alkalies 1789-1810. *Isis*, **65**, 59-65.

Le Grand, H.E. (1974b). Ideas on the composition of muriatic acid and their relevance to the oxygen theory of acidity. *Annals of science*, **31**, 213-25.

Levere, T.H. (1971). *Affinity and matter: Elements of chemical philosophy 1800-1865*. Oxford: Clarendon Press.

Levere, T.H. (1980). Humphry Davy, 'the sons of genius', and the idea of glory. In Forgan (1980), 33-58

Levere, T.H. (1981). *Poetry realized in nature: Samuel Taylor Coleridge and early nineteenth-century science*. Cambridge: Cambridge University Press.

[Marcet, J.H.] (1817). *Conversations on chemistry*. 5th edn., 2 vols., London: Longman.

Martineau, H. (1849-50). *The history of England during the thirty years' peace, 1816-1846*. 2 vols., London: Charles Knight.

Melhado, E.H. (1981). *Jacob Berzelius: The emergence of his chemical system*. Lychnos-Bibliotek no. 34, Stockholm: Almqvist and Wiksell International.

Morell, J.B. (1972). The chemist breeders: The research schools of Liebig and Thomas Thomson. *Ambix*, **19**, 1-46.

Murray, J. (1801). *Elements of chemistry*. 2 vols., Edinburgh: T. Maccliesh.

Murray, J. (1806). *A system of chemistry*. 4 vols., Edinburgh: Longman, Hurst, Rees & Orme.

Murray, J. (1811a). Observations and experiments on the nature of oxymuriatic acid. *Nicholson's* (ser. 2), **28**, 132-52.

Murray, J. (1811b). Observations and experiments on the alkaline metalloids. *Nicholson's* (ser. 2), **28**, 241-9.

Murray, J. (1811c). On the nature of oxymuriatic acid. *Nicholson's* (ser. 2), **28**, 294-310.

Murray, J. (1811d). Further observations and experiments on oxymuriatic acid. *Nicholson's* (ser. 2), **29**, 187-202.

Murray, J. (1811e). On the nature of oximuriatic gas. *Nicholson's* (ser. 2), **30**, 226-35.

Murray, J. (1812a). Experiments on muriatic acid gas. *Nicholson's* (ser. 2), **31**, 123-33.

Murray, J. (1812b). Experiments on the existence of water in muriate of ammonia. *Nicholson's* (ser. 2), **32**, 185-97.

Murray, J. (1813). On the existence of combined water in muriatic acid gas. *Nicholson's* (ser. 2), **34**, 264-76.

Murray, J. (1814). *Elements of chemistry*. 3rd edn., 3 vols., Edinburgh: John Anderson.

Murray, J. (1818). Experiments on muriatic acid gas, with observations on its chemical composition. *Transactions of the Royal Society of Edinburgh*, **8**, 287-328.

Murray, J. (1819). *System of chemistry*. 4th edn., 4 vols., Edinburgh: Francis Pillans.

Murray, J. & Murray, G. (1834). *A popular view of chemistry*. Edinburgh: John Anderson.

Nicholson, W. (1800-01). Account of the new electrical or galvanic apparatus of Sig. Alex. Volta, and experiments performed with the same. *Nicholson's* (ser. 1), **4**, 179-87.

Parkes, S. (1818). *The chemical catechism, with notes, illustrations and experiments*. 8th edn., London: Baldwin, Cradock and Joy.

Partington, J.R. (1961-70). *A history of chemistry*. London: Macmillan.

Pearson, G. (1797-98). Experiments and observations made with the view of ascertaining the nature of the gaz produced by passing electric discharges through water. *Nicholson's* (ser. 1), **1**, 241-8, 299-305, 349-55.

Peel, W. (1805). Untitled letters. *Philosophical magazine*, **21**, 279-80; **22**, 152-5.

Pinch, T. (1986). *Confronting nature: The sociology of solar neutrino detection*. Dordrecht: Reidel.

Priestley, J. (1778). *Miscellaneous observations relating to education*. Bath, R. Cruttwell for J. Johnson.

Priestley, J. (1802). Observations and experiments relating to the pile of Volta. *Nicholson's* (ser. 2), **1**, 198-204.

Russell, C.A. (1959). The electrochemical theory of Sir Humphry Davy, Part I: The voltaic pile and electrolysis, Part II: Electrical interpretations of chemistry. *Annals of science*, **15**, 1-13, 15-25.

Russell, C.A. (1963). The electrochemical theory of Sir Humphry Davy, Part III: The evidence of the Royal Institution manuscripts. *Annals of science*, **19**, 255-71.

Shapin, S. (1979). The politics of observation: Cerebral anatomy and social interests in Edinburgh phrenology disputes. In *On the margins of science: The social construction of rejected knowledge*, ed. R. Wallis, Sociological Review Monographs no. 27, pp. 139-78. Keele.

Shapin, S. (1984). Pump and circumstance: Robert Boyle's literary technology. *Social studies of science*, **14**, 481-520.

Shapin, S. & Schaffer, S. (1985). *Leviathan and the air-pump: Hobbes, Boyle and the experimental life.* Princeton: Princeton University Press.

Shinn, T. & Whitley, R., eds. (1985). *Expository science: Forms and functions of popularisation.* Sociology of the Sciences Yearbook no. 9, Dordrecht: Reidel.

Stansfield, D.A. (1984). *Thomas Beddoes, M.D. 1760-1808:Chemist, physician, democrat.* Dordrecht: Reidel.

Sutton, G. (1981). The politics of science in early napoleonic France: The case of the voltaic pile. *Historical studies in the physical sciences*, 11, 329-66.

Sylvester, C. (1804). Observations and experiments to elucidate the operation of the galvanic power. *Nicholson's* (ser. 2), **9**, 179-82.

Sylvester, C. (1806). Repetition of the experiment in which acids and alkali are produced in pure water by galvanism. *Nicholson's* (ser. 2), **15**, 50-2.

Thomson, J. (1832-59). *An account of the life, lectures and writings of William Cullen.* 2 vols., Edinburgh and London: Blackwood.

Thomson, T. (1814). Annual review of improvements in physical science. *Annals of philosophy*, **3**, 13-4.

Thomson, T. (1815). Annual review. *Annals of philosophy*, **5**, 12-4.

Thomson, T. (1816). Annual review. *Annals of philosophy*, **7**, 27-30.

Thomson, T. (1817). *A system of chemistry.* 5th edn., 4 vols., London: Baldwin, Cradock & Joy.

Thomson, T. (1818). Reports of Davy's paper 'On the fallacy of experiments in which water is said to have been formed by the decomposition of chlorine'. *Annals of philosophy*, **11**, 220; **12**, 450.

Treneer, A. (1963). *The mercurial chemist: A life of Humphry Davy.* London, Methuen.

Volta, A. (1800). On the electricity excited by the mere contact of conducting substances of different kinds. *Philosophical transactions*, 403-31.

Walker, A. (1807). *Analysis of a course of lectures in natural and experimental philosophy.* 14th edn., London: J. Barfield.

Whitley, R. (1985). Knowledge producers and knowledge acquirers: Popularisation as a relation between scientific fields and their publics. In Shinn & Whitley (1985), 3-28.

Wright, T. & Evans, R.H. (1851). *Historical and descriptive account of the caricatures of James Gillray.* London: Henry Bohn.

Yearley, S. (1985). Representing geology: Textual structures in the pedagogical presentation of science. In Shinn & Whitley (1985), 79-101.

Ziemacki, R.L. (1974). 'Humphry Davy and the conflict of traditions in early nineteenth-century British chemistry' (University of Cambridge Ph.D. thesis).

University of Wisconsin

JAN SAPP

THE NINE LIVES OF GREGOR MENDEL

Gregor Mendel's short treatise 'Experiments on Plant Hybrids' is one of the triumphs of the human mind. It does not simply announce the discovery of important facts by new methods of observation and experiment. Rather, in an act of highest creativity, it presents these facts in a conceptual scheme which gives them general meaning. Mendel's paper is not solely a historical document. It remains alive as a supreme example of scientific experimentation and profound penetration of data. It can give pleasure and provide insight to each new reader-and strengthen the exhilaration of being in the company of a great mind at every subsequent study (Stern & Sherwood, 1966: v).

There is no greater legend in the history of science than that of the experiments of Gregor Mendel. Three moments in this legend are extraordinary: first, how in the 1860s, Mendel discovered the laws governing the inheritance of individual characters; second, how the scientific world failed to recognise the monumental importance of these findings during his life-time; and third, the remarkable 'rediscovery' in 1900 of what later came to be called Mendelism. Thus, after an eclipse of some 35 years Mendel's experiments became universally hailed as providing a foundation for a chain of scientific research that culminated in the Darwinian evolutionary synthesis of the 1930s and '40s and in the spectacular accomplishments of modern molecular genetics. Eisely (1961: 211) summarized this legend beautifully when he wrote:

Mendel is a curious wraith in history. His associates, his followers, are all in the next century. That is when his influence began. Yet if we are to understand him and the way he rescued Darwinism itself from oblivion we must go the long way back to Brunn in Moravia and stand among the green peas in a quiet garden. Gregor Mendel had a strange fate: he was destined to live one life painfully in the flesh at Brunn and another, the intellectual life of which he dreamed, in the following century. His words, his calculations were to take a sudden belated flight out of the dark tomblike volumes and be written on hundreds of university blackboards, and go spinning through innumerable heads.

H. E. Le Grand (ed.), Experimental Inquiries, 137–166.

If Mendel and his experiments on peas were neglected for 35 years today they are alive and well and show no signs of dwindling in curiosity or significance. Since Mendel's 'vindication' at the turn of the century, more attention has been given to analyzing and commenting on his experiments than any other experiments in biology. Why is this? What is the power of these experiments? Here we meet with an apparent paradox. Although almost everyone agrees that these experiments are central to modern biology, there is no consensus about their exact significance. The great heap of literature addressing his motives, his experimental protocols, his own beliefs about heredity and evolution and the exact nature of his discovery remains largely incoherent. There are almost as many different interpretations as there are commentators.

The nine interpretations that will be briefly examined in the present study may be summarized as follows:

1. Mendel was a non-Darwinian. Although he was an evolutionist, he did not entirely agree with Darwin's views and set out to disprove them (Bateson, 1909).

2. Mendel was a good Darwinian. His experimental protocols and reported results can be explained on the assumption that he had no objections to Darwinian selection theory (Fisher, 1936).

3. Mendel was not directly concerned with evolution at all. He placed it on the back burner while he investigated the laws of inheritance (Gasking, 1959).

4. Mendel rejected evolutionary theory (Callender, 1988).

5. Mendel laid out the laws of inheritance which justifiably carry his name ('Standard View'; *e.g.*, Zirkle, 1951 & Mayr, 1982).

6. Mendel was no Mendelian. He was not trying to discover the laws of inheritance and several Mendelian principles are lacking in his papers (Callender, 1988; Brannigan, 1979; 1981; Olby, 1979).

7. Some of Mendel's data was falsified (Fisher, 1936).

8. None of Mendel's data was falsified (*e.g.*, Beadle, 1966; Dunn, 1965; Olby, 1966; Wright, 1966; Thoday, 1966; Mayr, 1982; Pilgrim, 1984; Edwards, 1987; Van Valen, 1987).

9. Mendel's reported experiments set out in his paper of 1866 are wholly fictitious (Bateson, 1909).

Most of these interpretations aimed to explain the long neglect of Mendel's experiments. In general, 'long neglect' accounts look for similarities between what became accepted as Mendelism during the earlier twentieth century and what Mendel wrote in 1866: a study of the transmission of individual traits, independent assortment of characters, statistical analysis of their transmission. In sum, they see the concepts and methods that later became the principle way of investigating the mechanism of inheritance and discerning the nature of hereditary variations. Simply put, the problem is this: if Mendel's experiments provided a foundation for genetics and evolutionary investigations, why was this not recognized in his day? Why, for example, was there not a meeting of the minds, so to speak, between Darwin and Mendel? This problem was first raised by geneticists, who offered a variety of reasons for the long neglect of Mendel's work. The explanations offered began with social considerations about the alleged obscurity of the journal in which Mendel published his results. It has been suggested that Mendel was professionally an outsider, an amateur, a monk, 'abbot of Brunn', and that anticlerical attitudes may have interfered in the proper evaluation of his claims (Dunn, 1965: 19).

Most commentators have emphasized conceptual reasons; they single out conflicting theories of heredity and competing research interests as being responsible for Mendel's neglect. But, here too the reasons given are often divergent and conflicting. Some have claimed that Mendel's work on the laws of inheritance was overshadowed by the attention given to larger questions concerning the mechanism of evolution with the appearance of *The Origin of Species* in 1859 (Bateson, 1909: 2; Dunn, 1965: 19). Others have argued that Mendel was trying to provide evidence for evolution and was therefore neglected by non-evolutionists (Fisher, 1936). On the other hand, it has often been claimed that Mendel's methodology was unorthodox. According to this argument, hybridists of Mendel's times investigated the transmission of 'species characters' to determine whether or not new species could be formed by hybrids. Mendel's approach differed; he investigated single character differences, not species differences. His statistical manner of analysis was at odds with the current ways of investigating hybrids (*e.g.*, Gasking, 1959; Dunn, 1965). Yet this interpretation, like all the others, has been contradicted. Long ago, the geneticist Conway Zirkle (1951) showed that Mendel's statistical approach was not unorthodox compared to existing traditions. In recent years, the view that Mendel's work was not 'ahead of his time' has been strengthened by the writings of historians and sociologists of science.

Studies by Callender (1988), Brannigan (1979; 1981) and Olby (1979) have challenged the 'long neglect' accounts of Mendel in a wholesale way. The principal conclusions of these investigations can be summarized briefly as follows. First, contrary to accepted opinion, Mendel was not trying to discover new laws of inheritance. He belonged to a tradition of hybridists who were examining the possibility that hybridization might be a source of evolution. That is, new species did not result from selection of small hereditary differences as Darwinians would have it; they were formed through the hybridization of existing species. The central question for Mendel and his fellow hybridists was whether or not hybrids were variable or constant (breed-true). If they were constant they might mark the beginning of new species. Mendel approached this problem with the conception of constant and independently transmitted characters. The laws of inheritance were of concern to him only so far as they bore on the question of the evolutionary role of hybrids. This program of nineteenth-century hybridists contrasts with that of geneticists at the turn of the century who were not interested in hybridization as a means of speciation, but used hybridization as a means to determine the nature of hereditary variability which in turn provided the fuel for evolution. So Mendel's problematic, the way he understood his work, was different from that of geneticists at the turn of the century (Brannigan, 1979: 81; Olby, 1979; Callender, 1988). Second, Mendel did not develop the concept of paired hereditary factors equivalent to the alleles of classical geneticists (Olby, 1979). Third, Mendel did not enunciate a 'law of segregation' which he thought might be applicable to all plant hybrids (Callender, 1988). These accounts then provide us with a radically different image of Mendel. He was not the lonely pioneer ahead of his contemporaries, someone who made an intellectual leap so great that its significance could not be understood by them, but rather someone whose work was firmly situated in the context of the mid-nineteenth-century research program on hybridization. Later, at the turn of the century, the meaning of Mendel's work was 'misinterpreted' by geneticists to produce the long-neglect legend.

Despite the new efforts to put Mendel and his work into his historical context, there still is no consensus about Mendel's intentions, what his historical context 'really' was. For example, Brannigan (1979: 448; 1981: 106) suggests that Mendel saw hybridization as a solution to the evolution of organic forms. Based on Zirkle's work, he (1979: 440; 1981: 105) remarked, 'If anything, Mendel's reputation was modest not because he was so radically out of line with his times but because his identity with his contemporaries was so complete!' Callender (1988: 72), who does not refer to Brannigan, tends to disagree. He argues forcefully that Mendel, 'was an opponent of the fundamental principle of evolution itself.' According to Callender, Mendel had adopted 'a sophisticated form of the doctrine of Special Creation as proposed by Linnaeus.' That is, he accepted the general fixity of species but acknowledged a limited number of cases in which new species had arisen through hybridization. On Callender's view, then, Mendel, would be a good scientific creationist! To support his claim, Callender (1988: 41) argued in part on the basis of what was missing in Mendel's papers: a concept of hereditary mutation. Moreover, Callender argues that Mendel had a record of misrepresenting the views of others in his papers and that this, combined with his opposition to evolutionary theory, 'are quite sufficient to account for the failure of his theories to make any significant impression on serious scientific opinion of his time.' Callender (1988: 73) claims that Olby, who tends to agree with his interpretation of Mendel's attitude towards evolution, distorts or misunderstood Callender's arguments.

One might throw one's hands up in despair: where is the truth? My objective here is not to try to find the key to unlock the mystery of Mendel's 'real' intentions. I am not, the reader may be relieved to know, going to provide a new 'definitive' reading of Mendel's work, offer still another reconstruction of his thought process or try to provide further detail of his place in the nineteenth century. To understand the significance of Mendel's experiments, such an approach would be fruitless. One important generality is already certain: Mendel's experiments were not held to be significant in his day. They are held to be significant only in the twentieth century. Moreover, as we shall see, Mendel's place in twentieth-century science is not determined by his writings of the 1860s. Indeed, in view of the diversity of the accounts about Mendel, it is reasonable to suppose that his writings do not even constrain the diverse interpretations offered. We have to look elsewhere to understand them. How, then, has this Austrian monk and his experiments on garden peas come to move so many people? The answer to this question too lies more in the stories written *about* Mendel and his experiments than in the stories written *by* Mendel. This is not to suggest that the discovery or 'Great Neglect' is 'more an artefact arising from the inadequacies of academic research than a genuine problem deriving from the actual course of historical development', as Callender (1988: 41) claims or that 'The Great Neglect' is a product of historians of science, not of scientific history (Callender, 1988: 72). On the contrary, as we shall see, these stories are a genuine problem derived from the actual course of the historical development of genetics.

1. MAKING A DISCOVERER

We begin our exploration of the significance of Mendel's experiments by a discussion of how the story about his discovery, neglect and rediscovery was invented. From whence did it originate? What were the contexts in which Mendel's contribution was raised to the status of a discovery? The geneticist Alexander Weinstein (1977) showed clearly that the belief that Mendel's work was virtually unknown before 1900 dates back to statements made at the turn of the century by the 'rediscoverers' of 'Mendel's laws', de Vries, Correns and Tschermak. Each insisted that they had read Mendel only after they had conducted their experiments and reached their own interpretations. He understands this as an attempt on their part to protect their priority. Each of the rediscoverers, Weinstein (361) argued, 'was anxious to have his work regarded as independent of the work of Mendel and of the other rediscoverers.' In fact, there is a widespread belief among commentators on Mendel's rediscovery that de Vries at first intended to suppress any reference to Mendel, but that his plans were interrupted when he found that Correns and Tschermak were going to refer to him (see Sturtevant, 1965: 27). This is based on de Vries's failure to mention Mendel when he first announced his discovery in a short abstract written in French. He mentioned Mendel only later in two longer papers, one in German and one in French where he remarked that it was '*trop beau pour son temps*' (see Weinstein, 1977).

Brannigan, a sociologist, took this suggestion further and argued that Correns, realizing that he had lost priority to de Vries, referred to Mendel's work as a strategy to minimize his loss and effectively to undermine the priority of de Vries' claim to the discovery. This suggestion is supported by Correns' reaction to de Vries' abstract in his paper of 1900 entitled 'G. Mendel's Law Concerning the Behaviour of Progeny of Varietal Hybrids', of which the opening paragraphs (Stern & Sherwood, 1966: 119-20) read as follows:

The latest publication of Hugo de Vries: 'Sur la loi de disjonction des hybrides,' which through the courtesy of the author reached me yesterday, prompts me to make the following statement:

In my hybridization experiments with varieties of maize and peas, I have come to the same results as de Vries, who experimented with varieties of many different varieties of plants, among them two varieties of maize. When I discovered the regularity of the phenomenon, and the explanation thereof -- to which I shall return presently -- the same thing happened to me which now seems to be happening to de Vries: I thought that I had found *something new. But then I convinced myself that the Abbot Gregor Mendel in Brunn, had, during the sixties, not only obtained the same result through extensive experiments with peas, which lasted for many years, as did de Vries and I, but had also given exactly the same explanation, as far as that was possible in 1866.* Today one has only to substitute 'egg cell' or 'egg nucleus' for 'germinal cell' or 'germinal vesicle' and perhaps 'generative nucleus' for 'pollen cell'. An identical result was obtained by Mendel in several experiments with *Phaseolus* and thus he suspected that the rules found might be applicable in many cases.

Mendel's paper, which although mentioned, is not properly appreciated in Focke's *Die Pflanzen-Mischlinge*, and which otherwise had hardly been noticed, is among the best that have ever been written about hybrids, in spite of some objections which one might raise with respect to matters of secondary importance, *e.g.* terminology.

At the time I did not consider it necessary to establish my priority for this 'rediscovery' by a preliminary note, but rather decided to continue the experiments further.

So, Brannigan (1979: 422-3) argued, 'Mendel's revival in 1900 took place in the context of a priority dispute between Correns and de Vries and that this dispute led scientists to overlook the original intent of the earlier research.' He further suggests that the labelling of the discovery as 'Mendel's laws' was a strategy to neutralize the dispute. 'This', he claims (1981: 94) 'is perhaps the single most important fact in the reification of Mendel as the founder of genetics.'

To Brannigan, the case of Mendel's 'rediscovery' is a good example of his social attributional model of discovery (1981) which he counterposes to mentalistic models. That is, instead of viewing discovery in terms of the creative genius of scientists, discovery should be treated as a process of social recognition which only later appears to be mentalistic or independent. Within this problematic, the great problems presented by scientific discovery are not simply who said, did, or 'found' something first, or how several scientists may almost simultaneously converge on a single theoretical model or technical procedure. The question is not how ideas come to mind, but how specific contributions come to be regarded as discoveries. As Brannigan (1979: 448) put it, 'A theory of discovery should concern itself *not* with determining what makes discoveries happen, but with what makes certain happenings discoveries.'

It would be difficult to disagree with the general thrust of Brannigan's view of discovery. But, should we accept his view that Mendel's laws and his representation as 'the founding father' of genetics are largely artifacts of a priority dispute between Correns and de Vries and that this dispute led scientists to 'overlook the original intent of the earlier research'? One might think this plausible, for scientists are often only concerned with those who precede them in so much as they see in past work elements of what they take to be the truth. Looking at the past, they often impose their own framework of understanding on the work in question irrespective of the intentions of the author. From this perspective all would agree that Mendel's work was superior to that of his contemporary hybridists. On this basis, for example, Zirkle, who did so much to show how Mendel's methodology was not as unorthodox as commonly assumed, still insisted that Mendel deserves the recognition he has received from geneticists (1951: 103):

> To conclude, we may be certain that Mendel was acquainted with the work of Knight, of Sageret and of Gaertner and probably also knew of Dzieron's hybrid ratio. In addition he had clues which led to the work of Seton and Goss. All of these contributions should have aided him in designing his experiments and have alerted him in what to look for. Of course his knowledge of this previous work would not detract from his own great accomplishments in the least. All of the earlier work together does not constitute Mendelism. Mendel's own experiments are so much more extensive and precise than those which went before that we are still justified in crediting him as the founder of a science.

Brannigan's claim that there was no interest in Mendel's real intentions is, however, questionable on several grounds. Mendel's intentions were addressed by William Bateson (1902; 1909), R.A. Fisher (1936) and many other scientists to the present day. The fact that scientists have shown such a remarkable interest in Mendel's 'true intentions' deserves explanation. It is first necessary to highlight a second difficulty with Brannigan's suggestion: it implies that Mendel's intentions would be obvious had anyone bothered to discern them. Callender (1988) makes similar assertions. Yet, despite the many attempts to reconstruct Mendel's thought process, there have been and continue to be different opinions of Mendel's 'real intentions'.

Mendel kept no diary and wrote little about himself (see for biographical information Iltis, 1932; Olby, 1985). He published only two papers (1866; 1870). The main source for reconstructing his thought process is his paper of 1866 entitled 'Experiments in Plant Hybrids', a concise transcript of two reports given at the Brunn Natural History Society in 1865. The whole story of the development of the new theory is usually claimed to be given in these 44 printed pages. Scientific papers are not diaries. But, is his own account to be taken at face value? The work of several historians and sociologists have shown that scientific papers commonly misrepresent the thought process that accompanied the work that is described. Scientific papers are often designed so as to give a veneer of objectivity and 'matter of factness' to published claims. This obscures the intentions and biases of the author and the process by which results are produced. Mendel's remarks concerning his experiments and the non-evolutionary views of the hybridist Gaertner illustrate the point. Conflicting interpretations of the following passage (in both the original German and English translations) have been offered (trans. W. Bateson in Sinnot, Dunn & Dobzhansky, 1958: 442-3):

> Gaertner, by the results of these transformation experiments, was led to oppose the opinion of those naturalists who dispute the stability of plant species and believe in a continuous evolution of vegetation. He perceives in the complete transformation of one species into another an indubitable proof that species are fixed within limits beyond which they cannot change. Although this opinion cannot be unconditionally accepted, we find on the other hand in Gaertner's experiments a noteworthy confirmation of that supposition regarding variability of cultivated plants which has already been expressed.

Does this statement mean that Mendel was an evolutionist or a non-evolutionist? R.A. Fisher (1936: 118) stated: 'It will be seen that Mendel expressly dissociates himself from Gaertner's opposition to evolution, pointing out on the one hand that Gaertner's own results are easily explained by the Mendelian theory of factors.' Similarly, Gavin de Beer (1964: 208) commented: 'This passage comes as near to the acceptance of the mutability of species as anyone could wish.' Yet, Callender (1988: 54) offers exactly the opposite interpretation: "If this statement is to be taken literally, as Mendel most assuredly intended it to be taken, then it says quite simply that he gave *conditional acceptance to the view, expressed by Gaertner, 'that species are fixed within limits beyond which they cannot change.*' Nothing could be clearer." But, surely, anything could be clearer. We do not have to decide here which interpretation is the correct one. It is enough to recognise at this point that Mendel's literary style, his attempts to sound objective in his evaluation of Gaertner's views, his use of double negatives, obscure his own intentions. It is not surprising that there is no consensus about the meaning Mendel gave to his own experimental work.

I do not suggest that the priority dispute between de Vries, Correns and Tschermak was not important for the initial recognition accorded to

Mendel. I protest only against any tendency to reduce the significance of Mendel's experiments and his place as the founder of genetics to being an artifact of a priority dispute which supposedly led scientists away from examining his intentions. There is much more than this underlying the value scientists have attributed to his work. We need not one cause for understanding his place in the history of genetics, but several. Mendel and the meaning of his experiments have come to be clothed in various social and intellectual guises.

Any understanding of the significance of Mendel's experiments in biology would have to recognise the importance of 'founding father mythologies' in the social and intellectual construction of science. The aggrandizement of past scientists through stories of their heroic insights may play an important role in defining and strengthening emergent research traditions. Forman (1969) has exposed various myths in scientists' accounts of the discovery of X-ray crystallography. He interpreted these myths as attempts to strengthen the tradition of X-ray crystallography by 'tracing it to a higher, better, more supernatural reality of initial events.' He argued (1969: 68) that "the traditional account may be regarded as a myth of origins, comparable to those which in 'primitive' societies recount the story of the original ancestor of a clan or tribe." Olby (1979) has suggested the same explanation for understanding the aggrandizement of Mendel in scientists' accounts of the origin of genetics. If this view be developed I suggest it would help us understand Mendel's prominent place in genetic discourse and culture.

At the most general level, Mendel's experiments are laden with morality. The long neglect theme, portraying the discoverer as a creative genius clothed in monastic virtues pursuing the truth undauntedly on the lonely frontiers of knowledge, unappreciated by his contemporaries, has been important in keeping Mendel's experiments alive. Brannigan himself, following Barber (1962), has argued that one reason why the Mendel case has been so poignant is because it has been presented as a tragedy which appeals to our sense of moral indignation. As Barber notes (1962: 540): 'The mere assertion that scientists themselves sometimes resist scientific discovery clashes, of course, with the stereotype of the scientist as the open-minded man.' Brannigan (1979: 453-4) compares Mendel's case to that of Galileo: 'In both cases, great contributions went unrewarded by the local communities. In other words, the suppression of Galileo by the Church and the apparent obscurity of Mendel elicit a common moral reaction over the patent injustice experienced by each.'

The element of morality is also embodied in another feature of the Mendel legend: the question of whether or not Mendel was honest in reporting his data as first raised by Fisher (1936). The claim that there was no deliberate falsification in Mendel's work has received a great deal of support from geneticists. At first glance suc a debate might seem trivial. Who really cares if Mendel fudged some data? After all he was right. However, once we consider the important cultural role of founding fathers in defining groups, the intentions and motives of the celebrated originator becomes extremely important. It is not surprising that the interest in whether or not Mendel deliberately faked some of his data was first brought to public attention at centennial celebrations of Mendel's paper and centennial symposiums of the genetics clan (see Dunn, 1965: 12; Iltis, 1966: 209; Olby, 1966; Thoday, 1966; Wright, 1966: 173-5; Beadle, 1967: 337-8).

The search for purity of motives in founding fathers is pervasive in the history of science. The reconstruction of the thought process of a creative enius has been central to the Darwin industry (Shapin & Barnes, 1979). What is a stake in this controversy is whether or not Darwin was in any way part of or responsible for the political and ideological uses of his theory. It is well known today that 'evolution had been invoked to support all sorts of political and ideological positions from the most reactionary to the most progressive' (Young, 1971: 185). Several writers have charged that Darwin was influenced by the socio-economic views of Malthus. Others argue that he was as much a social darwinist as his contemporaries who appealed to 'Nature' to legitimate their political views (Moore, 1986). As Shapin and Barnes (1979: 127) have pointed out, 'Darwin's defence' rests upon three assertions:

> The first is that of internal purity: Darwin's *intentions* and *motives* in writing the *Origin* were above reproach, and his personal beliefs in 1859 were innocent of 'ideological' taint. The second is purity of ancestry: influences upon the *Origin* were entirely wholesome and reputable, nothing 'ideological' was gleaned from Malthus. The third assertion is purity of germ-plasm: nothing outward could *properly* be deduced from the theory in the *Origin*; truth does not blend with error; insofaras truth was used to justify social Darwinism, it was misused.

They concluded (133) that 'Darwin's defence is far better staffed and funded than its opposition'. The more interesting question is: Why has the trial been conducted at all? They can only suggest (134) an anthropological explanation:

> The scientific discipline of evolutionary biology had its font and origin in the person of Charles Darwin and in the text of 1859. Darwin is a sacred totem by virtue of his 'foundership' of modern biology: science is sacred, so must Darwin and his Book be sacred; both must be protected from contaminationby the profane. As the author of the *Origin* he must himself be pure; his thought must be unmingled with worldly pollutionand incapable of satisfactorily blending or combining with the suspect formulations of social Darwinism. Thus, 'influences' from the 'profane' Malthus can only be the spiritual emanations of mathematics and genuine science, or nonessential stimuli or manners of speech. And implications for social Darwinism can only be misunderstandings.

Although this is a speculative suggestion, it has a great deal of merit in helping us to understand why the motives of so many founding fathers have been scrutinized.

It might not come as a surprise that the recent claims that Mendel was not the founder of genetics and that he did not make a major (though neglected) discovery have been dismissed by biologists. Despite the arguments of some non-scientists, scientists themselves still insist on portraying Mendel as a 'hidden genius' whose work was ignored until 1900. The evolutionist Ernst Mayr (1982: 713) dismisses the view that Mendel belongs in the hybridist tradition and insists that Mendel was a true evolutionist and that:

> As a student of Unger and of the problem of evolution, Mendel was concerned with single character differences and not, like the hybridizers, with species essence. To understand this fully is important for the interpretation of Mendel's work. It is totally misleading to say that Mendel's conceptual framework was that of the hybridizers. It is precisely the breaking away from the tradition of hybridizers that characterizes Mendel's thinking and constitutes one of his greatest contributions.

Mayr (1982: 717-8) recognises that 'Olby and others' are right that

> Mendel did not by a single stroke, create the whole modern theory of genetics. He did not have the theory of the gene, but neither did his rediscoverers.... However, Mendel's various discoveries (segregation, constant ratios, independent assortment of characters), combined with new insights acquired between 1865 and 1900, led, one is tempted to say automatically, to the theory quite legitimately called Mendelian.

To Mayr, (1982: 725) this 'by no means diminishes Mendel's greatness'. On the contrary, by showing that Mendel's theory was not fully complete, Mayr argues, the work of Olby and others only make it easier to understand why it was ignored for 34 years.

The cultural importance of founding fathers and scientists' so-called 'myths of origins', help us to understand why accounts of Mendel's discovery and neglect are repeated over and over. However, scientists' stories about the long neglect of Mendel and his pea experiments do not simply reappear over and over. They also change in such a way that the thoughts and motivations of Mendel are often altered; he is dressed and redressed in new colours of allegiance. We have yet to explain one of the most striking features of the Mendel literature: the very diversity of 'the long neglect' accounts. Any complete account of discovery and of Mendel's place in genetic culture would have to recognize that scientists' accounts of history play various important roles in their knowledge-making process. They surround experimental evidence and constitute part of the art of persuasion in science (see Sapp, 1986; 1987; 1990). The stories about Mendel's discovery and neglect vary and we need to know their specific rhetorical function in the constitution of scientific knowledge. For example, there is a similar story of discovery, neglect and rediscovery about Archibald Garrod and the origins of biochemical genetics. In this case, it seems to be clear that the story about the long neglect of Garrod was constructed by some geneticists in the 1950s to support the truth of a specific model of genic control. According to this myth, the 'truth' (a one-to-one relationship between genes and enzymes), like the discovery of the laws of inheritance, had been suggested several decades earlier but given an unfair hearing (see Sapp, 1990). Similar neglect accounts in which scientists are held to have been given an unfair hearing are pervasive in science. One of the most recent is that of the German-American geneticist, Richard Goldschmidt, constructed by S.J. Gould (see Dawkins, 1988: 81-2, 231-41).

What is often at stake in scientists' reconstructions of Mendel's thought process is a definition of the concepts and/or movements that can be legitimately associated with the genetics tradition. Throughout the twentieth century the significance of Mendelian genetics has changed. For example, the first generation of geneticists viewed Mendelism to be in direct conflict with Darwinian selection theory. By the 1930s, Mendelism was held to be compatible with Darwinian selection theory. No doubt the meaning of many experiments can be and is continually renewed as science proceeds. However, it is not just the meaning scientists place on Mendel's experiments that change with the development of Mendelian genetics, the inferences as to the meaning Mendel himself placed on his experiments also changes accordingly.

The very facts that Mendel published so little and that his motives are not transparent in these few papers, has helped his experiments to survive. Certainly, establishing the motives and intentions of scientists is a precarious business at the best of times but in few cases do we have to rely so much on logical reconstructions of a scientific paper as we do in the case of Mendel. Mendel's experiments thus become a flexible resource and, as a founding father whose intentions are so important, he is adaptable indeed. In a sense, he was lucky enough to please anyone who had an axe to grind. He and his experiments function as a source around which each new generation of geneticists constructs its social and intellectual world of legitimate associations. The meaning of his experiments, what he actually 'discovered', how the discovery was achieved, how and why it was ignored, have been interpreted to support a great variety of debates in the history of genetic research. The significance of Mendel's experiments lies in the diverse ways in which commentators have constructed stories about them and used them in their knowledge making. The strength of these stories, I suggest, changes according to the power relations in the field of genetics and evolutionary biology.

The history of genetics research in the twentieth century is marked by various struggles among competing groups over the direction of, and approaches to, biological research. It is marked by conflicts between experimentalists (geneticists) and non-experimentalists (naturalists and statisticians) over whether evolution is continuous or discontinuous and conflicts between experimentalists (embryologists and geneticists) over whether or not Mendelian genes controlled only superficial characteristics of the organism. We can find all these issues reflected in geneticists' accounts of Mendel's neglect. To follow scientists' reconstructions of Mendel's thought and the reasons for his neglect by the nineteenth century is to follow some of the central controversies in the development of genetic research in the twentieth century.

The idea that everyone saw in Mendel's experiments what they wanted to see is not a new interpretation. In fact, it was first suggested by Fisher (1936) in a noted paper, 'Has Mendel's Work Been Rediscovered?'. This was the first detailed attempt to reconstruct Mendel's thought process. By following the main lines of argument in Fisher's paper, we can begin to unravel the reasons for some of the diverse interpretations of Mendel's experiments and examine the wealth of issues that have kept them alive and vibrant. As we shall see, the specific representations of his experiments and intentions by biologists often reflect divergent and conflicting interests in the field of heredity and evolutionary theory. We begin with the way Fisher reconstructed Mendel to suit his own interests.

2. APPROPRIATING THE FOUNDING FATHER

Fisher's reconstruction came at a time when Darwinian theory was becoming intimately allied with Mendelian genetic principles. Fisher himself had played a leading role in this synthesis. In part, Fisher's paper represented an attempt to understand the famous dispute between the biometricians (statisticians) and the Mendelians (experimentalists). Conflicts between statisticians and Mendelian geneticists emerged at the turn of the century, soon after Mendel's laws were 're-discovered' (see Provine, 1971; Kevles, 1980; Mackenzie, 1981). The dispute revolved around the question of whether evolution was continuous or discontinuous: did new species emerge slowly by natural selection or quickly through large mutations, with natural selection in the latter playing only the negative role of eliminating new species or mutants which could not survive.

Biometricians, led by Karl Pearson and W.F.L. Weldon in England, saw themselves as Darwinians. They supported continuous evolution. Mendelians, led by William Bateson, were non-Darwinians and supported discontinuous evolution. The dispute raged in private correspondence and in published journals throughout the first decade of the century. Bateson found it 'impossible to believe' that biometricians had 'made an honest attempt to face the facts.' He doubted that they were 'acting in good faith as genuine seekers of the truth' (quoted in Kevles, 1980: 442). Weldon (1901) attempted to test the validity of Mendelism by subjecting Mendel's results to statistical tests. He did not claim that Mendel's results were statistically too good to be true, but doubted the possibility of reproducing Mendel's results with further pea experiments. Weldon concluded his critique with the remarks that Mendel was 'either a black liar or a wonderful man.' He remarked to Pearson, in 1901, 'If only one could know whether the whole thing is not a dammed lie!' (quoted in Kevles, 1980: 445).

But, there was more to the debate than a theoretical discussion about evolution. Both Mendelians and biometricians were struggling to dominate the field; both based their work on different methods as well as different theories. Methodological issues became a principal stake in this controversy. Were the methods of the experimentalist or those of the statistician more appropriate for biological; *i.e.*, evolutionary, problems? Geneticists, using experimentation as their polemical tool, attempted to exclude biometricians from the field by denying the legitimacy of purely statistical approaches to heredity and evolution. The views of Wilhelm Johannsen -- who provided the central terms of genetics: 'genotype', 'phenotype' and 'gene' -- were representative of those experimentalists who supported discontinuous evolution (1911: 130):

Certainly, medical and biological statisticians have in modern times been able to make elaborate statements of great interest for insurance purposes, for the 'eugenics-movement' and so on. But no profound insight into the biological problem of heredity can be gained on this basis.

Thus, non-Darwinian geneticists attempted to exclude Darwinian statisticians from the field. The dispute between the biometricians and Mendelians came to a head in England by 1905. Bateson was judged to be the victor (Provine, 1971).

However, by the 1920s and 1930s statisticians began to re-establish their authority in the field. They gained legitimacy mainly from statistical studies of populations spearheaded by Fisher, Haldane and Wright. They were the architects of what Julian Huxley in 1942 called the 'modern synthesis'. This synthesis was based upon Mendelian gene recombination, mutation and Darwinian selection theory. Evolution according to this heory was continuous: Bateson and the first generation of geneticists are judged to be wrong in allying Mendelism with non-Darwinian views of discontinuous evolution.

Fisher's paper of 1936 fits squarely within this theoretical shift. One can understand it as an attempt to drive the last nail in the coffin of the controversy. Central to his paper is a interpretation of Mendel's motives and theoretical views. The principal stake is a historical dispute over Mendel's attitude towards Darwinian natural selection. Bateson had cast Mendel as a non-Darwinian ally in his struggle against Darwinian biometricians. Fisher attempted to recast the founding father, Mendel, as a good Darwinian. Both Bateson and Fisher superimposed their own motivations and the context of their own work on Mendel.

The main thrust of Fisher's criticisms were aimed at Bateson. He used Bateson as a scapegoat for the heated controversy between Darwinians and Mendelians, charging him with intending to deceive scientists by allying Mendel and Mendelism with non-Darwinian views and by fabricating and distorting history to suit his interest (1936: 116):

It cannot be denied that Bateson's interest in the rediscovery was that of a zealous partisan. We must ascribe to him two elements in the legend which seem to have no other foundation: first, that Darwin's influence was responsible for the neglect of Mendel's work and of all experimentation with similar aims; second, that Mendel was hostile to Darwin's theories and fancied that his work controverted them.

As mentioned, Bateson and the first generation of Mendelian geneticists were in struggle with non-experimentalists, many of whom believed that Mendelism had little to do with the origin of species. Bateson, who actively promoted genetics, frequently mocked the integrity of alternative and conflicting approaches to the study of heredity and evolution. He claimed that alternative approaches and views of evolution were based on mere speculative theorizing which he believed stood in the way of sound experimental investigations of heredity. As Bateson (1914: 293) wrote:

> Naturalists may still be found expounding teleological systems which would have delighted Dr. Pangloss himself, but at the present time few are misled. The student of genetics knows that the time for the development of theory is not yet. He would rather stick to the seed pan and the incubator.

Appropriating Mendel, Bateson immediately began to tell stories in his scientific papers and books about Mendel's intentions and the neglect of Mendel's work. Bateson and Saunders (1902: 6) suggested that the principle of natural selection 'had almost completely distracted the minds of naturalists from the *practical* study of evolution. The labours of hybridists were believed to have led to confusion and inconsistency, and no one heeded them any more.' In *Mendel's Principles of Heredity*, Bateson claimed that like himself, Mendel had worked in virtual conflict with non-experimentalists and Darwinians and that this was partly responsible for Mendel's 'neglect' for 35 years. Thus Bateson (1909: 2) wrote:

> While the experimental study of the species problem was in full activity the Darwinian writings appeared. Evolution, from being an unsupported hypothesis, was at length shown to be so plainly deducible from ordinary experience that the reality of the process was no longer doubtful. With the triumph of the evolutionary idea curiosity as to the significance of specific differences was satisfied. The *Origin* was published in 1859. During the following decade, while the new views were on trial, the experimental breeders continued their work, but before 1870 the field was practically abandoned.

The suggestion that Darwin's influence was partly responsible for the neglect of Mendel's work was promoted by other geneticists who had pioneered the development of Mendelian analysis at a time when many biologists believed it was a minor curiosity with little bearing on the grand problems of evolution. For example, L.C. Dunn (1965: 18), who had been engaged in such polemics during the second decade of the century, tended to share Bateson's interpretation of Mendel's neglect:

> There is probably some truth to the explanation often offered, that Mendel was dealing with the minor tactics of evolution, and only indirectly at that, at a time when biologists had their thoughts and ambitions focused on the kind of grand strategy represented by the *Origin of Species.*

But, Bateson went further than Dunn. He not only suggested that Mendel was in virtual struggle with naturalists, he also imposed (1909: 311) a non-Darwinian motive on Mendel:

> With the views of Darwin which were at that time coming into prominence Mendel did not find himself in full agreement and he embarked on his experiments with peas, which as we know he continued for eight years.

'Had Mendel's work come into the hands of Darwin', Bateson (1909: 316) declared, 'it is not too much to say that the history of the development of evolutionary philosophy would have been very different from that which we have witnessed.'

Fisher strongly opposed Bateson's interpretation and claimed it was self-interested and held no truth value. He argued (1936: 117):

> Bateson's eagerness to exploit Mendel's discovery in his feud with the theory of Natural Selection shows itself again in his misrepresentation of Mendel's own views. Although he was in fact not among those responsible for the rediscovery, his advocacy created so strong an impression that he is still sometimes so credited.

Fisher, who 'knew' that *Mendelism* was not opposed to natural selection, believed that *Mendel* also knew that his work was allied with Darwinism. Those who believed that natural selection was the principle driving force in evolution could share both Mendel and Darwin as common intellectual ancestors.

When reconstructing Mendel's thought process, Fisher claimed that Mendel's experimental program could only be made intelligible on the basis that he worked squarely within a Darwinian framework. For example, Fisher claimed that in Mendel's day most hybridists crossed different species. They believed that species did not evolve and that they possessed essential qualities, specific natures or 'essences'. They were concerned with crosses between species to investigate the ways in which the forms of the hybrid reflected the parental essences. Mendel's approach, Fisher reasoned, conflicted with this: he crossed closely allied varieties not different species. This suggested to Fisher not only that Mendel was an evolutionist, but that his work was actually carried out within a Darwinian framework. Thus, Fisher (1936: 117) wrote: 'It's a consequence of Darwin's doctrine, that the nature of hereditary differences between species can be elucidated by studying heredity in crosses within species.' Fisher suggested that Mendel himself would have agreed with such views as evidenced, he claimed, by Mendel's crossing of varieties rather than species. Moreover, Fisher (1936: 118) argued, Mendel had claimed that his 'laws of inheritance' formed a necessary basis for understanding the evolutionary process. 'Had he considered that his results were in any degree antagonistic to the theory of selection it would have been easy for him to say this also.'

If this be the correct interpretation of Mendel's experimental program, then how did it go undetected for so long? Fisher (1936: 137) concluded his attempts to reconstruct Mendel as a good Darwinian by raising two issues in this regard. First, he claimed (like Brannigan and Callender later) that Mendel's opinions had been misrepresented because his work was not examined with sufficient care. Writers relied on accounts of others but, as Fisher remarked, 'there is no substitute for a careful, or even meticulous, examination of all original papers purporting to establish new facts.' Second, Fisher suggested that biologists before him had imposed their own meanings on the work of Mendel. Their interpretations were influenced by the theory of their times (1936: 137):

> Each generation, perhaps found in Mendel's paper only what it expected to find; in the first period a repetition of the hybridization results commonly reported, in the second a discovery in inheritance supposedly difficult to reconcile with continuous evolution. Each generation, therefore, ignored what did not confirm its own expectations.

The reading of scientific papers, like the construction of original scientific data, is not a straightforward affair. Meaning is not embedded in raw observations. It is bestowed upon the data by the observer. Nor is a unique, timeless meaning embedded in scientific papers reporting original data. It is often superimposed onto such papers. As Fisher suggests, when one reads a scientific paper one does so with theoretical expectations in mind. In his view, the biases of others were obstacles to the recognition of Mendel's discovery. However, once we recognize that all scientists have such biases, the issue of discovery is not one of unveiling certain truths which lay hidden in nature, or in past scientific papers. It is a matter of constructing the discovery.

Strikingly, Fisher excluded his own interpretation from any biases. Many of his claims have been challenged. It would not be difficult to show some of the ways in which Fisher himself shaped the evidence from Mendel's paper in order to impose a Darwinian framework on him. In effect, this was done by Gasking (1959). She was the first non-scientist to offer a detailed account dealing with the long-existing question: 'Why was Mendel's work ignored?' She (1959: 68) argued that although Mendel knew of the controversy surrounding the origin of species, his experiments were not directly concerned with them at all. Instead, she claimed (1959: 68) that 'he was from the outset looking for laws governing the inheritance of particular characteristics.'

Like Fisher, Gasking based her argument on a logical reconstruction of Mendel's experiments. She did not refer to Fisher's paper of 1936 and it is unlikely that she had read it. However, in effect, she debunked one of Fisher's arguments that Mendel's protocols can only be understood if he were a Darwinian. First, she pointed out that Mendel used both varieties and species in his experiments. Fisher was only telling a half-truth. According to Gasking (1959: 68), Mendel was simply 'indifferent whether his crosses were between species or only between varieties.' But, this was not because he was a proponent of Darwinian evolution, as Fisher suggested. In her view, 'Mendel's thinking was more like a farmer's than a biologist's.' She explained that unlike hybridists of Mendel's day who were concerned with 'specific natures' or 'essences' of a species, 'farmers and stock-breeders have a different problem' (1959: 61):

> They are concerned with not with the complete nature of a species, but rather with a particular property: they want cattle of larger size, beets with a higher sugar content, or whatever it may be, and the importance of inheritance for them lies in the results of crossing plants or animals having this particular property in different forms and degrees, Mendel's interest in inheritance was similar, and so differed fundamentally from that of other biologists.

Gasking's paper is representative of the 'long neglect' or 'rediscovery accounts' of Mendel's experiments, an approach that should be abandoned in favour of an 'anthropological' and sociological approach to understanding the power of Mendel's experiments. As I have suggested above, this alternative perspective helps us to understand still another central aspect of Mendel's experiments that has been so poignant among biologists: were Mendel's reported results faked?

3. RECONSTRUCTING MENDEL'S DATA

Since Fisher's 1936 paper, a great deal of attention has been given to the question of whether or not Mendel deliberately fudged his data. In view of Mendel's stature in genetic culture and the defence he subsequently received by geneticists, it might be questioned why Fisher, a Mendelian himself, would make such a charge in the first place. After all, fraud charges are often made to discredit an individual and/or competing theory. It was in the course of constructing Mendel as a good Darwinian that Fisher made the claim that Mendel's results were too good to be true and calculated that in the overall results one would expect a fit as good as Mendel reported once in 30,000 repetitions. However, this charge was not meant to discredit Mendel; it was meant to celebrate his power of abstract reasoning. Fisher (1936: 123) argued that Mendel had his laws in mind before he did his experiments:

> In 1930, as a result of a study of the development of Darwin's ideas, I pointed out that the modern genetic system, apart from such special features as dominance and linkage, could have been inferred by any abstract thinker in the middle of the nineteenth century if he were led to postulate that inheritance was particulate, that the germinal material was structural and that the contributions of the two parents were equivalent. I had no idea that Mendel had arrived at his discovery in this way. From an examination of Mendel's work it now appears not improbable that he did so and that his ready assumption of the equivalence of the gametes was a potent factor in leading him to his theory. In this way his experimental programme becomes intelligible as a carefully planned demonstration of his conclusions.

In Fisher's account, the claim that Mendel's data was too good to be true provides testimony to his claim that Mendel had his ideas in mind before doing his experiments. Mendel was a thinker not a tinker. But, did he cook his results to suit his theory? Fisher entertained three possibilities to account for Mendel's results: he was lucky; he unconsciously biased the results; he consciously biased the results in favour of his theory. Fisher ruled out the first two as providing inadequate accounts and favoured the third, 'fudging the data'. However, he did not hold Mendel responsible; rather, he suggested that possibly 'Mendel was deceived by an assistant who knew too well what was expected' (Fisher, 1936: 132).

Fisher's analysis of Mendel's data raises another set of issues for methodological reflection: this time about observer bias, the theory-ladenness of observations and whether or not the validity of experimental results could be tested statistically. Indeed, although Fisher's reconstruction of Mendel represented part of the process of closing the dispute between Mendelism and Darwinism, at the methodological level the conflict between statistical and experimental modes of reasoning continued. Geneticists who subsequently addressed Fisher's claims found it necessary to consider these methodological differences when attempting to understand the strength of statistical critiques of experimental results. Some of the difficulties to be encountered are well illustrated by a critique of Fisher's paper by the microbial geneticist George Beadle in the proceedings of the Mendel Centennial Symposium sponsored by the Genetics Society of America in 1965. Beadle charged that Fisher's reconstruction of Mendel's methods was incomplete and he explored the phenomenon of unconscious bias to account for Mendel's results. He claimed that Fisher had considered one kind of bias only, due to 'misclassification' of some hereditary variations; *e.g.*, 'a shrivelled round pea scored as unwrinkled'. Beadle remarked (1967: 338), 'As every experimenter in genetics knows, some classifications are difficult and may easily be unconsciously biased in favour of a preconceived hypothesis'. He was personally sensitive to this source of error, for as he recalled:

> I once discovered a loose genetic linkage in maize between floury endosperm and a second endosperm character known to be on chromosome 9, a linkage that I subsequently concluded was the result of my 'wanting' to find it. The floury character is often difficult to score, and I believe I unconsciously put the doubtful ones in the piles that would suggest linkage.

However, Beadle was careful to protect his own credibility and added, 'Fortunately, I recognized the possibility of this kind of error in time to withdraw a manuscript I had submitted for publication.'

Observer bias in selecting and sorting data is a serious obstacle for those who claim objective status for their experimental results. However, observer bias in selecting data in genetic analysis is only one difficulty. Beadle discussed a second problem resulting from the theory-ladenness of observations: how much data to include in a scientific paper and how an experimenter knows when the experiment is over. He suggested that it was entirely possible that Mendel stopped counting when he obtained results close to expectation. This possibility was also suggested by Dunn (1965) and Olby (1966). Beadle (1967: 338) explained:

> As he [Fisher] points out, Mendel clearly had his hypothesis in mind before completing all his work and therefore rejected certain numerical ratios. It is also clear, as Fisher deduces, that Mendel did not classify all the pea plants and seeds he grew. Presumably he classified enough to convince himself that the result was as expected. It is perfectly natural under these circumstances to keep running totals as counts are made. If, then, one stops when the ratio 'looks good', statistically the result will be biased in favour of the hypothesis. A seemingly 'bad' fit may be perfectly plausible statistically, but one may not think so and add more data to see if it improves, thereby raising interesting questions, some mathematical and some psychological.

What is of concern to us is *not* if Beadle's remarks actually account for Mendel's particular results, but the methodological issues they raise: data that looks good to the experimentalist looks bad for the statistician and *vice versa*. There seems to be a methodological incommensurability concerning the nature of statistical and experimental modes of reasoning. This might be called the 'experimentalist-statistician paradox': from a statistical point of view, geneticists should not provide 'too much data' and have their results come too close to theoretical expectations, for the closer they come to the 'truth' the less true they will appear to be. This strange paradox is based on faulty reasoning. The reason why data are considered to be less true the closer they reach theoretical expectations is based on the idea that geneticists *should be* studying a random sample. It assumes that experiments should be carried out independently of the law or theory the observer is using for explanation. In other words, it appeals to naive empiricism and ignores the theory-ladenness of observations. The theory itself informs the experimenter about what kind of experiment to perform, what kind of phenomena to examine and how results are to be understood; it also tells the experimenter when the experiment is over. This last issue is at the heart of Beadle's suggestion that Mendel simply stopped counting when he obtained the results expected.

What liberties scientists are 'allowed' to take in selecting positive data and omitting conflicting or 'messy' data from their reports is not defined by any timeless method. It is a matter of negotiation. It is acquired socially: scientists make judgments about what fellow scientists might expect in the way of methods, data and standards, in order to be convinced. What counts as good evidence may be more or less well-defined after a new speciality is formed but at revolutionary stages in science, when new theories and techniques are being put forward and standards have yet to be negotiated, scientists have less an idea of what others may expect to be competent and convincing. Statistical criticisms were weak and could be trivialized in as much as they ignored aspects of the experimental process: the conscious and unconscious biases of geneticists in selecting certain phenomena to investigate and certain data to report. One could not evaluate the validity of Mendel's experimental claims on statistical grounds alone. Those such as Sewall Wright (1966: 173-4) who were accomplished in both statistics and genetic experimentation recognized the limitations of statistical criticisms:

I do not think that Fisher allows enough for the cumulative effect on X^2 of a slight subconscious tendency to favour the expected result in making tallies. Mendel was the first to count segregants at all. It is rather too much to expect that he would be aware of the precautions now known to be necessary for completely objective data....Checking of counts that one does not like, but not of others, can lead to systematic bias toward agreement. I doubt whether there are many geneticists even now whose data, if extensive, would stand up wholly satisfactorily under the X^2 test.

4. THE RHETORICAL NATURE OF SCIENTIFIC PAPERS

Fisher's paper of 1936 forces us to examine another aspect of the experimental process -- the extent to which published experimental reports can be taken as literal accounts of how scientists generate and interpret their data. This issue was raised by Bateson and Fisher when attempting to understand Mendel's conduct and determine the liberties he may take. Contrary to general belief, Fisher was not the first to question the authenticity of Mendel's reported results. Although ignored by commentators who have examined Fisher's statistical criticisms of Mendel, Bateson's comments raised the possibility that all of Mendel's 'experiments' were fictitious. He (1909: 350) questioned the authenticity of Mendel's experiments in a footnote to a passage in the translation of Mendel's experiments he used in his book, *Mendel's Principles*. Mendel, after describing his first seven experiments, opened his subsequent section with the following claim: 'In the experiments described above plants were used which differed only in one essential character.' Bateson commented:

This statement of Mendel's in the light of present knowledge is open to some misconception. Though his work makes it evident that such varieties may exist, it is very unlikely that Mend l could have had seven pairs of varieties such that the members of each pair differed from each other in *only* one considerable character.

Fisher fully realized the weight of this criticism. One would expect that some or all of the crosses would have involved more than one contrasting pair of characters. Fisher believed that Mendel meant his reports to be taken literally. In response to Bateson's remarks, he offered two possibilities to account for Mendel's statement. Both involved how Mendel wrote up his reports and what he regarded as an 'experiment'. The first possibility was that: 'He might, for each cross, have chosen arbitrarily one factor, for which that particular cross was regarded as an experiment, and ignored the other factors' (1936: 119). Although this way of analyzing crosses might seem to be wasteful of data, Fisher claimed that Mendel, in fact, 'left uncounted, or at least unpublished, far more material than appears in his paper.' In other words, he published only enough data that he believed sufficient to convince readers of his theory. The second possibility was that: 'He might have scored each progeny in all the factors segregating, assembled the data for each factor from the different crosses in which it was involved, and reported the results for each factor as a single experiment' (1936: 119). This, Fisher claimed, is what most geneticists would take, unless they were discussing either linkage or multifactorial interaction.

On the other hand, Bateson's intimation that Mendel's 'experiments' were fictitious remained a possibility. As Fisher noted, Mendel did not give summaries of the aggregate frequencies from different experiments. This conduct would be easily intelligible if the 'experiments' reported in the paper were fictitious, being in reality themselves such summaries. This kind of over-simplification is often used when teachers illustrate principles to students in a lecture. Fisher (1936: 119) continued:

> Mendel's paper is, as has been frequently noted, a model in respect of the order and lucidity with which the successive relevant facts are presented, and such orderly presentation would be much facilitated had the author felt himself at liberty to ignore the particular crosses and years to which the plants contributing to any special result might belong. Mendel was an experienced and successful teacher, and might have adopted a style of presentation suitable for the lecture-room without feeling under any obligation to complicate his story by unessential details. The style of presentation with its conventional simplifications, represents, as is well known a tradition far more ancient among scientific writers than the more literary narratives in which experiments are now habitually presented. Models of the former would certainly be more readily accessible to Mendel than of the later.

It is difficult to know exactly what Fisher meant by the tradition of 'ancient' scientific writers. However, one can easily challenge any sharp distinction between what Fisher called the 'simplifications' of 'ancient' scientific writers and those 'more literary' accounts of modern scientists.

Peter Medawar in 1963 posed the question: 'Is the Scientific Paper a Fraud?' Medawar did not mean that the scientific paper misrepresents 'facts', nor that the interpretations found in a scientific paper 'are wrong or deliberately mistaken'. What he meant was that 'the scientific paper may be a fraud because it misrepresents the thought process that accompanied or gave rise to the work that is described in the paper' (1963: 377). Medawar was perhaps the first to emphasize that 'The scientific paper in its orthodox form *does* embody a totally mistaken conception, even a travesty, of the nature of scientific thought.' The structure of the 'orthodox scientific paper' itself Medawar argued, is telling in this regard. He described the structure of the typical scientific paper in the biological sciences as follows:

> First, there's a section called the 'introduction'in which you merely describe the general field in which your scientific talents are going to be exercised, followed by a section called 'previous work' in which you concede, more or less graciously, that others have dimly groped towards the fundamental truths that you are now about to expound. Then a section on 'methods' - that's O.K. Then comes the section called 'results'. The section called 'results' consists of a stream of factual information in which it's considered extremely bad form to discuss the significance of the results you're getting. You have to pretend that your mind is, so to speak, a virgin receptacle, an empty vessel, for informationwhich floods into it from the external world for no reason which you yourself have revealed. You reserve all appraisal of the scientific evidence until the 'discussion' section, and in the discussion you adopt the ludicrous pretence of asking yourself if the information you've collected actually means anything; of asking yourself if any general truths are going to emerge from the contemplationof all the evidence you brandished in the section called 'results'.

The above description is somewhat of an exaggeration, for certainly many scientific papers do not follow this structure., but we can agree with Medawar that there is 'more than a mere element of truth in it.' 'The conception underlying this style of scientific writing is that scientific discovery is an *inductive* process' (1963: 377). In its crudest form, induction implies that scientific discovery, or the formulation of scientific theory, begins with the 'neutral' evidence of the senses. The scientific paper gives the illusion that discovery begins with simple unbiased, unprejudiced, naive and innocent observation. Out of this unbridled evidence and tabulation of facts orderly generalizations emerge, crystallize or at least gel. Yet scientists know full well that discoveries do not emerge and gel in this way. They know what meaning to place on their results before they conduct their experiments. Indeed, it is their anticipation of results that informs them of what experiments to perform, what phenomena to examine and what data to report. Medawar traces the inductive structure often framing modern scientific papers to the nineteenth century writing of the philosopher John Stuart Mill. However, it would be naive to believe that scientists are the dupes of philosophers. It is also wrong to suggest that 'the scientific paper' is a fraud. 'The scientific paper' is not a fraud; it is rhetoric. The structure of the narrative of the scientific paper plays an important persuasive role in science. First, it is important to remember that scientific work is steeped in the biases of two cultures: the larger culture in which science is allowed to persist and the scientific culture itself.

Scientists' belief in theories, the experiments and observations they make and report are often influenced by forces arising from both. In the larger culture scientists have maintained their legitimacy in part by appealing to their 'objectivity'. The literary style of the scientific paper is designed to protect scientists' interests as purveyors of truth and to maintain public support. The structure of the scientific paper plays a similar rhetorical role within the scientific culture. Shapin (1984) gives a detailed study of an early attempt to construct conventions for writing scientific papers. He shows that Robert Boyle set out rules to distinguish authenticated scientific knowledge from mere belief. This was done, in part, by what Shapin calls 'the literary technology of virtual witnessing.' This '*literary technology* by means of which the phenomena produced...were made known to those who were not direct witnesses', involved providing protocols for experiments, recounting unsuccessful experiments and displaying humility so as not to look self-interested and untrustworthy, citing other writers not as judges but as witnesses to attest matters of fact, *etc*. Boyle's literary technics would give a veneer of objectivity and 'matter of factness' to published scientific claims. But, this often obscures the intentions of the author and the process by which results are produced. Indeed, the method sections of scientific papers are often imprecise. Medawar's deconstruction of 'the scientific paper' is incomplete in this regard. Seemingly trivial but yet vital information concerning procedures is often left out of scientific papers (*e.g.*, Collins, 1985). As a result, as the case of Mendel illustrates, interpretations of a scientist's conduct and procedures may involve considerable speculation and conjecture. When commenting on Mendel's paper one writer remarked. 'All geneticists admitted that it was written so perfectly that we could not -- not even at present -- put it down more properly' (Nemec, 1965: 13). Yet, it was this very 'perfection' that has made Mendel's conduct so difficult to ascertain.

5. CONCLUSION

Mendel's 'Experiments on Plant Hybrids' has been shrouded in various myths about individual discovery and social neglect. The aim is not to debunk these myths and dismiss them, but to reveal them, study them and understand how they have been constructed, how they have persisted and how they have been altered since the turn of the century. The prominent place of Mendel's experiments in scientific culture is based on the strengths of these myths, the very diversity of the reconstructions of his thought process and his role as the founding father of genetics. As with all founding fathers, reconstructing the experiment has been closely intertwined with reconstructing the intentions of the experimenter. Mendel has been cast as an ideal scientific type, wrapped in monastic and vocational virtues. Yet, the moral element of the

Mendel legend is not the only thing that has captured scientists' interests. There is much more to the oft-repeated accounts of Mendel's neglect and the reconstruction of his thought process than a construction of an exemplar of scientific virtue, a representation of a scientific ideal. To discard the stories about Mendel's discovery and subsequent neglect simply as moral tales would be to ignore the important rhetorical role of Mendel's experiments in the construction of scientific knowledge.

Since the emergence of genetics, Mendel has become a cultural resource to assert the truth about not only what it means to be a good scientist, a geneticist, but also what Mendelian genetics implies. The divergent accounts of Mendel's neglect reflect the often conflicting social and intellectual interests of Mendel's commentators. To understand geneticists' reconstructions of Mendel's intentions is to understand the divergent and sometimes conflicting definitions of what Mendelian genetics signifies or connotes. The specificity of the accounts themselves are generated as part of the repertoire of rhetorical tools scientists have at their disposal when defending their social and intellectual positions in science. Geneticists' reconstructions of Mendel's true intentions are used to buttress conflicting claims about what concepts can be legitimately associated with Mendelism (continuous or discontinuous evolution, for example). We have seen how Bateson and Fisher constructed Mendel and the reasons for his neglect to suit their own intellectual struggles over evolutionary theory. And, this same pattern has been repeated over and over. Representing the foundations of genetics, Mendel's experimental results are used by geneticists to discuss what is legitimate experimental practice, to reflect upon the unconscious biases of experimentalists and to consider the procedures by which experimental claims can be evaluated. In short, Mendel's experiments are a meeting place where scientists discuss the definition of science itself.

REFERENCES

Barber, B. (1962). Resistance by scientists to scientific discovery. In *The sociology of science*, ed. B. Barber & W. Hirsch, pp. 539-56. New York: The Free Press.

Bateson, W. (1909). *Mendel's principles of heredity*. Cambridge: Cambridge University Press.

Bateson, W. (1914). Address of the President of the British Association for the Advancement of Science. *Science*, **40**, 287-302.

Bateson, W. & Saunders, E.R. (1902). Experimental studies in the physiology of heredity. *Reports to the Evolution Committee of the Royal Society*, i, pp. 1-160.

Beadle, G.W. (1967). Mendelism, 1965. In *Heritage from Mendel*, ed. R.A. Brink & E.D. Styles, pp. 335-50. Madison: University of Wisconsin Press.
Brannigan, A. (1981). *The social basis of scientific discoveries*. Cambridge: Cambridge University Press.
Brannigan, A. (1979). The reification of Gregor Mendel. *Social studies of science*, **9**, 423-54.
Callender, L.A. (1988). Gregor Mendel: An opponent of descent with modification. *History of science*, **26**, 41-75.
Campbell, M. (1976). Explanations of Mendel's results. *Centaurus*, **20**, 159-74.
Collins, H. (1985). *Changing order: Replication and induction in scientific practice*. London: Sage.
Dawkins, R. (1986). *The blind watchmaker*. Harmondsworth: Penguin Books.
De Beer, G. (1964). Mendel, Darwin and Fisher (1865-1965). *Notes and records of the Royal Society*, **19**, 192-226.
Dunn, L.C. (1965). *A short history of genetics*. New York: McGraw-Hill.
Edwards, A.W.F. (1987). Defending Mendel merely perpetuating a myth. *Nature*, **326**, 449.
Eisley, Loren. (1961). *Darwin's century*. New York: Anchor Books.
Fisher, R.A. (1936). Has Mendel's work been rediscovered? *Annals of science*, **1**, 115-37.
Forman, P. (1969). The discovery of the diffraction of X-rays by crystals: A critique of myths. *Archive for the history of exact sciences*, **5**, 38-71.
Gasking, E.B. (1959). Why was Mendel's work ignored? *Journal of the history of ideas*, **20**, 60-84.
I tis, H. (1966). *The life of Mendel*. New York: W.W. Norton.
Johannsen, W. (1911). The genotype conception of heredity. *American naturalist*, **45**, 129-59.
Kevles, D. (1980). Genetics in the United States and Great Britain, 1890-1930: A review with speculations. *Isis*, **71**, 441-55.
MacKenzie, D. (1981). *Statistics in Britain, 1865-1930*. Edinburgh: Edinburgh University Press.
Mayr, E. (1982). *The growth of biological thought*. Cambridge: Harvard University Press.
Medawar, P. (1963). Is the scientific paper a fraud? *The listener*, 12 Sept., pp. 377-8.
Moore, J. (1986). Socializing darwinism: Historiography and the fortunes of a phrase. In *Science as Politics*, ed. L. Levidow, pp. 38-80. London: Free Association Books.
Nemec, B. (1965). Before Mendel. In *Fundamenta genetica*, ed. J. Krizenecky, pp. 1-13. Osterhout and Brno.
Olby, R. (1966). *Origins of mendelism*. New York: Schocken Books.
Olby, R. (1979). Mendel no mendelian? *History of science*, **17**, 53-72.
Pilgrim, I. (1984). The too good to be true paradox and Gregor Mendel. *The journal of heredity*, **75**, 501-2.
Provine, W.B. (1971). *The origins of theoretical population genetics*. Chicago: University of Chicago Press.
Sapp, J. (1986). Inside the cell: Genetic methodology and the case of the cytoplasm. In *The Politics and Rhetoric of Scientific Method*, ed. J. Schuster & R. Yeo, pp. 167-202. Dordrecht: Reidel.
Sapp, J. (1987). *Beyond the gene: Cytoplasmic inheritance and the struggle for authority in genetics*. New York: Oxford University Press.
Sapp, J. (1990). *Where the truth lies: Franz Moewus and the origins of molecular biology*. New York: Cambridge University Press.

Shapin, S. (1984). Pump and circumstance: Robert Boyle's literary technology. *Social studies of science*, **14**, 481-520.

Shapin, S. & Barnes, B. (1979). Darwin and social darwinism: Purity and history. In *Natural order: Historical studies of scientific culture*, ed. S. Shapin & B. Barnes, pp. 125-39. Beverly Hills: Sage.

Sinnot, E., Dunn, L.C. & Dobzhansky, T. (1958). *Principles of genetics*. New York: McGraw-Hill.

Stern, C. & Sherwood, E.R., eds. (1966). *The origin of genetics: A Mendel source book.* San Francisco: W.H. Freeman.

Sturtevant, A.H. (1965). *A short history of genetics*. New York: Harper & Row.

Thoday, J.M. (1966). Mendel's work as an introduction to genetics. *Advancement in science*, **23**, 120-34.

Van Valen, L.M. (1987). Mendel was no fraud. *Nature*, **325**, 395.

Weinstein, A. (1977). How unknownwas Mendel's paper? *Journal of the history of biology*, **10**, 341-64.

Wright, S. (1966). Mendel's ratios. In *Origin of genetics*, ed C. Stern & E.R. Sherwood, pp. 173-5. San Fransisco: W.H. Freeman.

Young, R. (1971). Darwin's metaphor: Does nature select? *The monist*, **55**, 442-503.

Zirkle, C. (1951). Gregor Mendel and his precursors. *Isis*, **42**, 97-104.

University of Melbourne

DAVID TURNBULL & TERRY STOKES

MANIPULABLE SYSTEMS AND LABORATORY STRATEGIES IN A BIOMEDICAL INSTITUTE

1. PROLOGOMENA

In recent years, there has been a wide, interdisciplinary focus on experiment in science (Batens & Van Bendegem, 1988; Collins, 1985; Galison, 1987; Hacking, 1983; Knorr-Cetina, 1981; Knorr-Cetina & Mulkay, 1983; Latour & Woolgar, 1979; Pinch, 1986; Rouse, 1987; Shapin & Schaffer, 1985). All are agreed in portraying the laboratory as a site where an activity central to science occurs. From the perspective of the sociology of scientific knowledge it is in the laboratory that scientists, instead of passively discovering reality, actively manufacture knowledge. Scientific experiment, rather than being in Reichenbach's (1951: 97) phrase, simply 'a question addressed to nature', is seen as the opportunity for scientists through a variety of social, literary and technical practices to accomplish 'facticity', 'truth', 'objectivity' and 'efficacy' (Latour & Woolgar, 1979, 180-3).

In the popular view -- perhaps as much as in that of scientists themselves -- there has always been an ambiguity about whether the essence of science is experiment or theory. Philosophers of science, with their epistemological preoccupations, have almost invariably seen theory as the essential core of science. Thus Agassi (1988: 4), notes:

> According to the traditional view, experiments are designed to put theories to the test.... [Thus] experiments depend *genetically* and *logically* on theories, since they are conceived and designed as explicit questions formulated by means of the concepts of a theory and with a view to testing it [emphasis in the original].

However, this orientation led to the recognition that the relationship of experiment to theory is fraught with difficulties. The theory-ladenness of observation, underdetermination of theory by data, and ad hoc auxiliary hypotheses render extremely problematic the warrantability of experimental knowledge claims and have led to the abandonment of the notion of an *experimentum crucis*; that is, to the abandonment of the

H. E. Le Grand (ed.), Experimental Inquiries, 167–192.

idea that nature can definitively cry 'Yes!' or 'No!' to any of our claims about it (Duhem, 1954).

These philosophical critiques provided the conditions for the possibility of the sociology of scientific knowledge (Barnes, 1974; Bloor, 1976; Mulkay, 1979a). But, possibly reflecting these epistemological roots, some sociology of scientific knowledge has, in our view, taken an overly 'linguistic turn' (Mulkay, Potter & Yearley, 1983; Gilbert & Mulkay, 1984; Mulkay 1985). While it was important to emphasize the social construction of knowledge claims; inevitably, indeed deliberately, this was at the expense of materiality. Thus Collins (1985: 16) advocates treating 'descriptive language as if it were about imaginary objects.' The consequence has been, as Rouse (1987: 26) puts it, to portray science as a 'network of theory rather than a field of practices.'

An adequate sociology of scientific knowledge must certainly recognize that scientific knowledge does not follow as a matter of necessity from some privileged conjunction of method, experiment and theory, and is hence a social process. But, it must also recognize that it is the result of the application of human skills in particular local and contingent circumstances. Thus, while requiring theorizing and methodical attentiveness, science is at root, both a social and a practical process; a process whose indeterminate and contingent character produces outcomes that could be other than they are. That science could be 'other' and is rooted in human action is the most important insight that has emerged from the sociology of scientific knowledge and is essential to an emancipatory and critical understnding of science.

We argue that the inclusion of material practice is not incompatible with the inherent relativism of the sociology of scientific knowledge position (Turnbull, 1984). Rather by adopting a form of materialism that Pickering (1985) has called pragmatic realism, the sociology of scientific knowledge can be materialist, relativistic and reflexive. Though there are no fixed absolute criteria external to human practices that enable us to adjudicate whether a particular causal explanation is true of reality, we do have a body of accepted beliefs and practices that enable us to produce desired effects in particular circumstances. Those beliefs and practices and the criteria by which we evaluate them are constitutively social but they gain their 'truth' for us, through active human engagement with the material world, that is - among other things, through experiment. This point is well made by Ian Hacking (1983: 208-9) in his discussion of the way we come to believe in the existence of entities that are not directly observable like electrons or cell structures:

We are convinced not by a high powered deductive theory about the cell -- there is none -- but because of a large number of interlocking low level generalizations that enable us to control and create phenomena in the microscope. In short we learn to move around in the microscopic world.

Hacking argues that we learn about the world through intervening and that as analysts we should focus on practice as well as theory or representation.

Ten years ago Michael Mulkay argued that a sociological analysis which leads to to the recognition that 'the procedures and conclusions of science are like cultural products, the contingent outcome of interpretive social acts (1979b: 65)' is not contradicted by arguments based on the practical effectiveness of science, but he left for further research the question of how practical effectiveness is itself socially constructed. In their focus on interpretative flexibility (Collins, 1985) and the problematic relationship between representations and objects, sociologists of science have left unexamined the source of science's material power, of its practical efficacy. We suggest that the peculiar power and efficacy of science lies in the materiality which conditions scientific knowledge as the result of the labour involved in its production (Young, 1979; 1985; MacKenzie, 1981).

2. MANIPULABLE SYSTEMS

We want to consider scientific research as a labour process. This requires focussing on human agency, social organization, experimental apparatus and raw materials. Karen Knorr-Cetina (1981) pointed out that scientists are basically concerned with 'getting things to work.' The first problem for the scientist is to get the experimental set up to produce interpretable results. Only after that comes the struggle over interpretation. Once the apparatus is deemed to be working the focus of debate is no longer on the data as such but on representations of it in various interpretative contexts (Ackermann, 1985: 33). Of course it is not the instruments alone that produce the data, but instruments, skills, techniques and raw materials, together with theories, all embedded in social relations. It this combination of social and material factors that we call a 'manipulable system'. It is, we will argue, the establishment and maintenance of such manipulable systems that is the primary task of the experimental scientist. We have used the term 'system' to emphasize the interrelatedness of the constituent elements and the heterogeneous engineering required to produce and maintain such systems (Hughes, 1985; Law, 1987).

An effective manipulable system gains its power from enabling two orders of manipulation. The first involves manipulation of the otherwise undifferentiated phenomena of nature to render them observable, demarcatable, measurable and recordable; *i.e.*, to generate data. This process reverses the scales of magnitude and time. The invisible and rare are made large and frequent, the large and unencompassable are made small and locally variable. The fast are made slow and the slow speeded up. Eventually everything from epidemics to particles is processed, tamed, captured and recorded. Representations of data, unlike the phenomena themselves, are capable of being 'read', superimposed, synthesized, integrated and transmitted, hence they enable the second order of manipulation (see Latour, 1986). Precisely because the successful establishment of such a system provides scientists with greatly enhanced power in both generating and manipulating data, the system itself has very significant effects on structuring and guiding research.

To show how the concept of a manipulable system illuminates the material practice of science, we consider some of the work we did as part of an extended ethnographic study of the Walter and Eliza Hall Institute of Medical Research, in Melbourne (Charlesworth *et al.*, 1989). This institute, known as the Hall Institute or WEHI, provides ample opportunity for observing such manipulable systems in action, since it is itself so structured as to provide for their ready formation. After briefly considering the larger structure we will focus on two groups of scientists at the WEHI: one group trying to produce a malaria vaccine; the other working on thymocyte differentiation.

3. THE WEHI AS GENERATOR OF MANIPULABLE SYSTEMS

WEHI, established in 1915, is Australia's oldest specialized biomedical research establishment (Charlesworth *et al.*, 1989; Burnet, 1971). By 1987, total staff was in the vicinity of 300, working in a brand new \$A40 million facility with an annual expenditure exceeded \$A14 million (WEHIMR, 1987: 3). Since the late 1950s, its research has centred on immunology. F.M. Burnet, the immediate past Director, shared a Nobel Prize for his work on immunological tolerance and is best known for his clonal selection theory of the immune response. The WEHI is associated, and indeed physically connected with the Royal Melbourne Hospital, and to the adjacent University of Melbourne, where it constitutes the Department of Medical Biology. But, unlike a hospital, WEHI research includes but is not dictated by clinical problems. Unlike an ordinary university department, it does not have an inflexible structure determined by undergraduate teaching requirements.

Like many biomedical research institutes, WEHI has an agenda which ranges from applied to pure research, which is primarily driven by considerations of scientific significance and which is generally justified by potential application to human disease only in the medium to long run. Of critical importance is structural flexibility - the ability to expand, contract or create new research areas as opportunities arise. For much of Burnet's term as Director, WEHI was largely virological in orientation but, as his interests turned to immunology, WEHI's focus also changed. The current Director, G.J.V. Nossal, has also attempted to ensure both flexibility and specialization. He himself has been responsible for major developments in molecular immunology (taking early advantage of recombinant DNA technology) and, most recently, immunoparasitology (mainly to try to develop such vaccines as an anti-malarial). His conscious strategy is to maintain a structure which enables opportunistic flexibility within the broad range of immunology. To do this he has concentrated on establishing the essential preconditions for generating successful manipulable systems: trained personnel, awareness of trends through international connections, adequate funding, raw materials supply and a flexible structure.

WEHI devotes much effort to training technicians, graduate students and post-doctoral fellows. It attracts some of the best Australian scientific and medical graduates, usually placing them in overseas fellowships when they attain their doctorates. There is also steady foreign interest in doctoral and postdoctoral work at WEHI. Thus inbreeding is avoided and international communication maintained. Similarly, most senior scientists constantly travel to foreign conferences and laboratories. There is a programme of sabbatical leave and international collaboration is common (indeed, more common than collaboration with Australian workers in other places). Many serve on such international bodies as the World Health Organization. For 1986-89 Nossal was President of the International Union of Immunological Societies. It is equally important to ensure a supply of skilled personnel who are abreast of the latest techniques. Technicians are especially important here, since it is they who do much of the bench work. Secure and flexible funding is obviously another essential and, though WEHI has a substantial block grant from the National Health and Medical Research Council (indeed, the largest such grant in Australia), it also insists that all researchers constantly seek funding of their own. The Institute's endowment is around $A18 million and there is an active programme to increase it.

Rather less obvious but nonetheless extremely important is the maintenance of a guaranteed supply of raw materials and equipment. To this end WEHI maintains at considerable effort and expense two animal breeding facilities - one providing a totally specific-pathogen-free environment - that assure a standardized supply of healthy mice in dozens of varieties (with currently unused varieties kept as frozen embryos). WEHI's research agenda is very heavily structured by this major raw material: So much so that WEHI could be said to 'run on mice'. Over 100,000 are 'sacrificed' each year. They are of course very special mice: disease-free and genetically pure. All are laboratory artefacts: some, for example, lack an immune system. In contrast, the Howard Florey Institute across the road runs on sheep and other large domestic species. The respective research agendas of WEHI and the Florey are structured so markedly by reliance on different animals that despite a common concern with human diseases there is little interaction other than in recombinant DNA methodologies.

Structurally, the key element in the WEHI's organization are the Units, made up of groupings of research laboratories. These are serviced and supported by specialized and centralized facilities for electron microscopy and histology, protein structure, cell counting and typing, media preparation and glassware sterilization, radiation safety and instrument standardization, computing, library artwork and photography, engineering, catering, word processing, purchasing and stores. There are also financial and administrative sections. These auxiliary services profoundly influence the rate at which data can be generated by manipulable systems. The WEHI's research structure, centred on Units, facilitates growth in some areas and contraction in others and thus provides a high degree of flexibility in responding to changes in research foci (Studer & Chubin, 1980: 14). The rise of immunoparasitology well illustrates this.

4. SETTING UP THE 'MALARIA SYSTEM'

In 1987 the smallest Unit, Thymus Biology, had barely a dozen scientific and support staff whereas the largest, Parasitology, had more than sixty (WEHIMR, 1987). The first spawned the second. During 1975, Graham Mitchell set up an Applied Immunobiology Laboratory within what was then the Experimental Pathology Unit (WEHIMR, 1976). He had previously worked with the Unit Head, J.F.A.P. Miller, on B-cell/T-cell collaboration and, when he returned from an overseas post-doctoral stint, was encouraged by the Director to set up his own laboratory. Mitchell's initial intention was to investigate the immunology of foetal/maternal relations. The area was interesting because it was not understood why the maternal immune system does not reject the foetus as 'other' and funding was available. However, foetal/maternal relations was dropped as a research area in favour of host/parasite systems because, as Mitchell says, 'we just didn't have the techniques, just didn't have the way of approaching the question.' Or, as we would put it, he was unable to establish a manipulable system.

With the switch to host/parasite studies, the name of Mitchell's lab was changed to the Laboratory of Immunoparasitology (WEHIMR, 1978). Within four years, Mitchell had become Head of a new Immunoparasitology Unit with two laboratories - his own Immunoparasitology and a Molecular Parasitology Laboratory run by Robin Anders. Miller's Experimental Pathology Unit was renamed the Thymus Biology Unit (WEHIMR, 1982). The Immunoparasitology Unit continued to grow: the work on malaria attracted another senior researcher, David Kemp, from the Molecular Biology Unit to work within that Unit (WEHIMR, 1984). Malaria research burgeoned still further with support from the MacArthur Foundation. Kemp became head of a new MacArthur Laboratory of Molecular Parasitology (WEHIMR, 1985). The Unit itself was then renamed the Parasitology Unit (WEHIMR, 1987).

When Mitchell began work in the late 1970s only two parasites were available to the laboratory for studies in mice: *Ascaris suum* and *Babesia rodhaini*. But, within eighteen months there were four people working on setting up manipulable parasite systems. Karen Forsyth, now at the Institute of Medical Research in Papua New Guinea (IMRPNG), recalls:

> We weren't talking about malaria vaccines or anything in those days, we were concerned with just simply addressing the question of how one understands the host parasite relationship and particularly the relationship between the immune system and the parasite. We were hampered by a number of problems, in particular the lack of parasite material we could work with. So we were very concerned about that problem, it seemed to be a major hurdle to get over, before we could even go on to identify what were protective antigens and therefore what one could consider candidate antigens for which ever disease. We all knew that one approach would be to get into molecular biology and start cloning. I think it was not until cloning was achieved that people started thinking of vaccine. We looked at a number of systems and I had no feeling that 70% of the Unit's effort would end up on malaria. We had leishmaniasis, intestinal nematodes, malaria, tapeworms; then the Rockefeller money came and it was the big push into human malaria. We were just spreading the net wide and trying just to see which systems were

goers.... In the early days we were not thinking in terms of a malaria vaccine at the end of ten years work at the Unit. It could have been a vaccine for anything, it was what went well, basically.

Initially, then, the research agenda was set by the politics of international funding by the World Health Organization (WHO) and the Rockefeller Foundation, the strategy of proliferating parasite systems, and the constraints of raw materials, skills and techniques (Charlesworth *et al.*, 1989: ch. 10). Eventually, ten different parasite systems were established. However, as research money came in from WHO and the Rockefeller Foundation's Tropical Disease Research and the Great Neglected Diseases programmes, malaria became the major research focus. The parasite systems were cut back to *Babesia*, malaria, schistosomiasis, Taenia, Leishmania and Cysticerocis, all of which involved significant outside collaboration. Now, with increased emphasis on medical parasitology, they are working only on malaria, schistosomiasis and Leishmaniasis.

For the malaria researchers, the task was to establish a system for isolating antigens (proteins which provoke an immune response) from malaria parasites. There were a number of material constraints. For example, *Plasmodium falciparum* was the only one of the four species of protozoan parasite that cause human malaria that has ever been cultured in the laboratory (Trager & Jensen, 1976). Though *falciparum* is the most deadly, *Plasmodium vivax* accounts for 80% of all cases but it has not yet been cultivated *in vitro*. *In vitro* cultivation is a pre-requisite for an immunological approach to malaria: it is more manipulable for the purposes of assessing effects of antibodies and the labeling and isolation of antigens.

Having set about looking for *falciparum* antigens, some strategic decisions were necessary. Many antigens are stage-specific, so one decision was the stage in the parasite's life cycle on which they should concentrate. Several American groups with long-standing involvement in malaria research had already done much work on the sporozoite stage. This is the stage in which *Plasmodium* is injected into the blood stream by the *Anopheles* mosquito. However, it is only very briefly present in that form before it disappears into the liver, eventually to re-emerge, much multiplied, as a merozoite. Since a single sporozoite is sufficient to induce malaria, any vaccine for that stage would have to be 100% effective. Given this and the fact that the American group were at work on the sporozoite stage, the WEHI group opted for the asexual blood stage - the merozoite. Mosquitos are a prerequisite for work on the sporozoite stage and they then had neither the facilities or the expertise at WEHI to maintain colonies of mosquitos. This basic element of their research agenda -- what aspect of the parasite on which to work -- was set by a conjunction of the nature of the research problem, a social strategy reflecting the structure of the field at the time and the resources and skills they possessed.

The group now had to make another choice between strategic options: to try to pick the which antigens were likely to be functionally significant, or try to characterize and produce as many antigens as possible and then see which provoke an immunological response which might confer protection. WEHI scientists chose the second route, partly because so little was known about the biology and immunology of *falciparum* malaria. That choice posed two further immediate problems: how to obtain the antigens and select the significant ones. WEHI established collaboration with the Papua New Guinea Institute of Medical Research (IMRPNG) which had developed a well-documented, malaria-affected human population in Madang and were able to supply the essential raw materials: blood with malaria parasites from infected patients and blood with antibodies from immune patients. This was an important resource for WEHI researchers, enabling them to work on large amounts of material containing current parasites from a defined population.

One difficulty in developing an effective vaccine is that many of the large number of proteins exhibited by the malaria parasite provoke antibody responses but most of the antigens confer no immunity. The easy answer to the question of how to pick the significant antigens would be to try them all and see. The ideal subjects for such a study are monkeys but the species required is in very short supply and those few available are held by the Centre for Disease Control in Atlanta. The significance of this is explained by Ross Coppel, one of the WEHI scientists working on the malaria project: 'If there was no shortage of monkeys we would essentially take every clone we have, purify out the malarial antigen, stick it into a monkey and test it. That would be it, there wouldn't be any underlying theory.' The essential problem, then, was how to establish a system to generate large amounts of antigenic material and to test it for it's immunological significance. Setting up this system and getting it to run effectively was no mean feat. It took nearly two years of hard and relatively unproductive work to achieve what the WEHI researchers thought of as 'the breakthrough':

> It was very different then..people worked longer hours there were a lot of difficult experiments, unsuccessful experiments and a lot of slogging and an enormous amount of pressure from the Director. He'd say 'What the hell is going on? Why isn't it working,

hasn't something happened yet?' There was a lot of push but very little productivity. The difficult things in science are ha d to achieve and they are a slog. It is just literally sitting there and trying everything you can think of until it works. When cloning worked out there was a great celebration and it was really like the technical breakthrough had been made. Once that breakthrough had been made then clearly unlimited experiments could be done and when a scientist reaches that point, that's the point of celebration. Its not, 'Yes I can get a vaccine next year', it's 'I now have plenty of material, I can do as many experiments as I can think of.' The technique giving rise to lots of material; that was the achievement to me [Karen Forsyth].

The technical task was to get a bacterium to produce a protein which might prove useful for a vaccine. Mitchell reportedly said 'Until now the single most important impediment to the development of parasite vaccines has been the scarcity of parasite protein antigens available for testing (Westmore, 1983: 760).' At this point no recombinant molecular biology (rDNA) had been done on parasites. The key breakthrough was to use rDNA techniques to make *E. coli* express parasite proteins. However, cloning the antigens was only half the problem. As Kemp, Head of the Molecular Parasitology Laboratory observes, the other half was identifying the significant ones:

Screening the colonies with human antibodies was the way to go. First we improved the technique for doing this early in 1981. Then we had two years of making libraries and screening them and trying to soup up the assays. There was no indication for those two years that the screening would work. We were always mucking around with a new vector, a new assay, a new library, whatever. Then we got a new vector from Stanford which we modified and made a library which worked. Meanwhile we had developed better immunochemical screening system, we've been running round with thousands of screens ever since.

The vector that finally enabled the system to become effective was developed by one of the molecular biologists who brought it with him when he joined the WEHI group. That a genetically engineered tool for the transport of rDNA to a host cell can both be a piece of personal property and make the difference between success and failure exemplifies the local and contingent character of research and the constraints imposed by the availability of materials, techniques and skilled personnel (Turnbull, 1989).

The WEHI's manipulable system for the generation and identification of malaria antigens is characterized by the following sequential elements. First, the parasites from blood of PNG malaria patients are cultivated in the laboratory. Second, the antigens are extracted and their genes cloned. Since each antigen is a protein that has a gene the encodes it, the section of DNA for that gene is identified and spliced into the bacterium *E. coli*, colonies of which will then express it. The result is a pure colony expressing a single parasite gene that encodes an protein. Third, antibodies are identified and extracted from the blood sera of people in PNG, in particular those who provide clinical evidence of resistance to malaria. The antigens of significance being those that are differentially recognized by resistant and susceptible individuals. Fourth, the antigens which provoke immunologically effective antibodies are identified by finding those that bind to antibodies which have been radioactively labeled. Ideally all the antigens should be selected, but in a library of 20-30,000 parasite proteins only a proportion of the probable 300 antigens are obtained and only four or five of them might provide resistance. Finally, those antigens that correlate with induction and expression of host protective immunity are selected as candidate vaccine antigens. By 1986 the WEHI malaria parasite manipulable system had produced a four-part candidate antigen which was successfully tested for its capacity to induce resistance in monkeys at the Centre for Disease Control in Atlanta.

5. TECHNIQUES, INSTRUMENTATION AND EQUIPMENT

The establishment of the parasite manipulable system depended critically on techniques. The system was jointly dependent on the techniques of molecular biology and more traditional parasitology. To the innovation of culturing parasites *in vitro* were added rDNA techniques for identifying genes of interest and splicing them into a suitable bacterial host, which expresses them, together with the techniques for using antibodies (including monoclonal antibodies, whereby a cell producing a specific antibody is fused with an immortal cancer cell, yielding hybridoma colonies which can also be cultured *in vitro*) to detect expressing clones. Some of these techniques were once the exclusive skills of molecular biologists in whose discipline they originated but now they have become pervasive throughout the WEHI and biomedical research generally. The usefulness of the tool kit of the molecular biologist in addressing a plethora of questions, including immunological ones, has enabled a whole new set of manipulable systems to be developed. Just as tissue culture techniques had earlier freed experimenters from reliance on patients and animal models, so, monoclonal antibodies extended the kinds of *in vitro* experiment which could be done. Molecular biological techniques could be readily deployed even by those relatively unfamiliar with the theory and problems of the field to which they were being applied. The molecular biologist, Dave Kemp, comments:

> The prospect of moving into malaria research was agreeable 'scientifically'. I'd be in a position to guide the technology without worrying about what the questions were. I didn't know a thing about parasites. Becoming a self perpetuating system is what happens- good money, good people, good work, and so it goes.

That a manipulable system is as much structured by instrumentation as it is by technique and materials is apparent in the following comments by Ross Coppel, who describes the malaria research as 'hand to mouth science':

> If you think about, say, the pulse field work, it was really 'Here's a new machine, now let's play around with what it does.' There has been a lot of, sort of tinkering and playing, and out of it has come some useful results. But the grand overlying theory, even as far as the rational planning of experiments for malariology goes, just hasn't been helpful in the work we've done. For example, we'd very much like to clone the cytoadherence molecule. So you come out with a rational set of strategies to maximize the chance of finding a clone pattern for the molecule, but since half the things we really need for that, we don't have, like ready access to parasites growing in monkeys; we just sort of try by trial and error to find them. So even when you have got the plan and theory and strategy, you may not be able to put it into place. There's been a very simple sort of reasoning going on. We have just simply done experiments, something's popped out and its been very obvious what the next step has been. There haven't been any great subtle insights or much of that sort going on.

It is also the case that much of the data produced by the WEHI malaria group is dependent on instrumentation. Though much of the apparatus is commercially available and is therefore in a sense universal, many instruments and their associated techniques are either produced in-house or given a local modification. The automatic protein sequencer temporarily installed in Melbourne University as a joint facility for WEHI and the Melbourne branch of the Ludwig Cancer Institute is a case in point. The availability of the protein sequencer (which determines the sequence of amino acids in a polypeptide chain) has altered the range of research that can be undertaken. For example, once a protein is isolated and sequenced, it is possible to determine the corresponding DNA sequence (within limits, owing to redundancy in the DNA code). As we will show in a later section, the Fluorescence Activated Cell Sorter (FACS), on which much of the work on white blood cell development is dependent, provides a particularly clear example of 'technology-driven', as distinct from 'technique-driven' research.

6. THEORIES

As we talked to the WEHI scientists and observed them at work, we were looking for the dynamic which drives and structures their activities. The traditional view is that scientific research is theory-driven (Agassi, 1988). Theories propose explanations and make predictions which are to be confirmed or disconfirmed experimentally; lacunae suggest fruitful areas of experimental investigation. Bhaskar (1983: 138), says: 'Theory without experiment is empty. Experiment without theory is blind. But the precise role of experiment in relation to theory remains a matter of philosophical (and historiographical) dispute.' Physicists have long identified themselves as theoreticians or experimentalists and the social structure within which they work reflects this. It is quite unremarkable for a scientist to be described as a theoretical physicist - meaning one who is mainly or wholly engaged in theoretical work, doing little or no experimental work, and probably located in a Department of Theoretical Physics. Many scientists at the WEHI, however, see theory as playing a secondary role, rather than providing the main dynamic. Some immunologists there, for example, describe their discipline as 'soft' or 'very forgiving.' By this they mean that the generally accepted macroscopic theoretical framework --clonal selection theory -- does not impose very rigorous constraints on, or provide guidance for day-to-day experimentation. Macroscopic theory does not readily generate potential solutions to micro-level problems. It is unlikely that an insouciance to theory is special to molecular parasitologists, but it may be peculiar to biology and biomedicine. Even those biologists who do have a penchant for theorizing rarely do only that.

Macfarlane Burnet is best known for his theoretical contributions - most notably immunological tolerance (for which he shared the 1960 Nobel Prize for Physiology and Medicine with B.P. Medawar) and the clonal selection theory of the immune response. Yet, Burnet regarded himself as first and foremost an experimentalist. When he summarized his scientific achievements, he listed the main topics of his experimental work first (1968: 4-6), in keeping with his claim to have 'enjoyed the life of the experimental scientist sometimes almost to the pitch of ecstasy (1968: 246).' He was ambivalent about the role of theory. On the one hand, he wrote (1968: 213):

I am sympathetic with the attitude of the modern professionally trained research-worker, that the production of theories and generalizations should be restricted to areas where the facts and the experimental techniques available already allow a definite conclusion to be reached. No particular credit attaches to finding the right answer by a guess, inspired or not, made at a time when the facts necessary to give the answer were not available.

Yet, in the succeeding paragraph he continues:

That is one attitude. Like most of my generation, amateurs compared with our successors, I take a different viewpoint. I believe that scholarly work in science must aim at generalization and that research lacks meaning if it is not at every stage guided and stimulated by hypotheses, including major ones, devised on the best available data and analogies and put in such a form that in principle an experimental discussion, proof, disproof or modification, should be eventually possible.

A biomedical ambivalence toward theory can be seen in remarks of Ross Coppel, a WEHI malaria vaccine researcher with a background in molecular biology and medicine:

Theory has actually been surprisingly small in the role it has played. In fact the molecular biology turned the project a bit on its head. We had a very rational strategy when we first started. We started off with getting the biology going and then the immunochemistry so that we could identify some antigens which would be important and clone them. Then RESA got plucked out of the blue, a clone picked at random because it just happened to make a large fused polypeptide which would be easy to work with. An antigen which has never been previously described and we weren't expecting to find. Something on the surface of an [immature] ring stage [of the malaria parasite] that had never been noticed before that turns out to be a very useful vaccine candidate. So in fact the theory all followed behind....You see, it's mainly because the systems themselves are so messy, and the data is open to so many interpretations that people's belief in a true state of affairs - in, say, invasion of the red blood cell by the [malaria] parasite are so waffly that they are not helpful. You cannot use them as a basis for experimental strategy.

Immunoparasitologists could, with some degree of rational expectation, set out to produce a vaccine for malaria, without any fully articulated theory to guide them. On the one hand there is sufficient macro-theory about rDNA and protein structure to enable the isolation of given molecules and on the other sufficient micro-theory about antibody response to specify the broad class of potentially significant proteins. However both the macro and the micro theories have unresolved difficulties and there is only the most tenuous of links between them. There is for example, as Graham Mitchell acknowledges, 'a major deficiency in our knowledge of the immunology of malaria, the basis of resistance in humans is still unknown'. Indeed he allows that 'some might say that this is why the malaria vaccine will take much longer to develop than some have anticipated or budgeted for.'

Despite the absence of fully articulated theory and rules, the power and efficacy of a manipulable system enabled the malaria research group to generate usable data. Indeed it was not until the manipulable system was fully operational that theory was able to play an overt role in structuring the research:

The hypothesis about cross recognition, that came out of the data. Now we are actually going about doing the traditional sort of science in that we now use theory and are going out to test that and see whether its going to make predictions about how the immune response should occur in the presence of one set of repeats or multiple repeats and we'll now do traditional testing of hypothesis and see if it works. If not modify and so on. But it didn't play much role till now. We didn't predict the existence of repeats. Finding them was a big surprise and now all our theories are trying to tie together observed experimental facts. To fit them together to make a nice story that would explain what you see. But the sort of structure of, hypothesis followed by theories, to get us through every hour every day wasn't really a very important factor. It was just sort of pressing ahead, doing the experiment, finding the result and then moving on from there [Ross Coppel].

Graham Mitchell, on the other hand, believes 'most of us do science by generating a few insights or hypotheses on the basis of the data which are tested using the appropriate technology, which leads to their modification.' This process "in actuality is 50% bread and butter, 30% new technological development and 20% "crazy experiments.""

Much of the research agenda seems to be structured by the material constraints of the manipulable system and to be relatively unaffected by theoretical concerns but this is largely a function of the incorporation of much of the theory in the techniques and instrumentation. As Karen Forsyth, the WEHI malaria group's PNG collaborator pointed out:

You can't look at one successful experiment. The fact is that someone's tried fifty other things that didn't work before that. Like the guy who cloned the circumsporozite protein. When he got a test tube of monoclonal antibodies from Ruth Nussenzweig he got ten years of biology saying this was the thing to clone.

The isolation of the so called RESA molecule is an indicator of the material efficacy of manipulable systems. One member of the malaria team described the presentation of results on the same molecule by their Swedish rivals:

What was presented at the Kyoto International Immunology meeting in 1983 by the Stockholm group was fluorescence on ring stages, the identificationof an antigen on ring stages. They were looking for ways of making infected red cells the targets of antibody dependent cytotoxicity. Then they were throwing in human sera and monitoring the system by doing immunofluorescenceand they found when they gluteraldahyde fixed they got this fluorescence on ring stages. No one had previouslyidentified or even thought that there was an antigen in the membrane of cells infected with ring stages, ...meanwhile we'd cloned them and found the same thing.

This is significant in that neither group had any reason to expect a RESA type antigen to exist but it was thrown up by two different manipulable systems. There is of course much rivalry between the teams as to who isolated it first and what it should be called, but for our purposes it is not necessary to assume that RESA exists, or that it is a surface antigen, or that it has any significant function in the immunology of malaria. It is sufficient to point out that it was produced independent of any theoretical expectation and it was produced by two independent research groups. The occurrence of the RESA molecule within two at least partially independent networks can be seen as a consequence of the material character of its production.

7. A MACHINE-DRIVEN LABORATORY

White blood cells, lymphocytes, are of two major sorts, B-cells and T-cells, both originating in the bone marrow. The T-cells migrate to the thymus, where they mature into several different types. The work of the Lymphocyte Differentiation Unit (LDU) focuses on this process. As such it is engaged in quintessentially basic, fundamental or 'pure' research - justified, in the customary way, by the claim that if we are to understand the pathological, we must first understand the normal. This provides one contrast with the applied vaccine work of the immunoparasitologists. Another is size: the whole LDU is about the size of any one of the three Laboratories composing the Molecular Parasitology Unit. The LDU has no formal internal structural divisions. Informally there are two groups, one led by the Unit Head, Ken Shortman, the other by Roland Scollay. Both groups do collaborative as well as independent work in closely related areas.

The LDU's work relies heavily on the WEHI Flow Systems Laboratory and its Fluorescence Activated Cell Sorter (FACS). The FACS machine passes cells in a stream only one cell wide through two or more different-coloured laser beams. With aid of a computer, the resulting patterns of scattered light and fluorescence are analysed so that particular kinds of cell - for example types of thymocyte (T-cells) - can be identified, counted and sorted into separate receptacles. The FACS facility was first installed in 1977 and has been continuously expanded, modified and upgraded since (WEHIMR, 1985: 20). Other researchers, including immunoparasitologists, make regular use of it but the whole research programme of the LDU is centred on the FACS.

Facilities like the FACS and the Protein Structure Laboratory shape research agendas in a variety of ways. Some (generally cheaper) tools of analysis like optical microscopes and balances or even small cell counters are idle part of the time but not the FACS. Projects in Units other than the LDU are designed around it and it is in constant use. As the capabilities of the machine are explored, its limitations are discovered. This generates a demand for expansion and development. Thus, in the course of 1988, two new FACS were ordered and would be integrated with existing machines and the Institute's computing facilities. Nevertheless, the temptations of interpretations in terms of technological determinism are to be avoided, for there are always local and contingent factors, some deliberate and some accidental. At one time there was in Scollay's laboratory a cell counter to which a microcomputer had been attached, thereby automating the process. But, it had been unused for more than a year for want of someone who knew how to operate the computer!

Demand for the FACS machine throughout the Institute is such that time on it must be booked well in advance. All other experimental activity has therefore to be carefully planned so that thymocyte populations are ready for counting and sorting when the lab has time on the FACS. Despite the expansion of the facility to accommodate a complete new flow cytometry system, dubbed WEHI FLOSY (WEHIMR, 1985: 20), dependency on strictly rationed FACS time has begun to become a disadvantage for Ph.D. students. Accordingly, Shortman and Scollay are advising new students to tackle projects which do not require the sophisticated central cell-sorter.

8. 'EXPERIMENT' IN ITS MICRO-SOCIAL CONTEXT

Just as an effective manipulable system is both dependent on and necessitates the local availability of specially tailored raw materials and instrumentation, so, too, does it necessitate and permit co-ordinated team research carrying out a multiplicity of experiments. In order to develop a malaria va cine the Molecular Parasitology Unit has had to establish co-ordinated research linking several laboratories. The work of each laboratory, each group within the laboratory and each individual within a group, is related to all the others both deliberately and accidentally. Thus the success, failure or delay in one sub-project advances or retards progress in another. In addition serendipitous results from one group can re-orient both its own and other groups' work. Such highly co-ordinated, large-scale, goal-directed research is still rather uncommon. In most biomedical laboratories, the concurrent projects are not closely related (other than by general subject matter). This poses its own problems, because projects sometimes pull at cross purposes. The laboratory head must constantly choose which to give priority.

Bigger laboratories often appoint a laboratory manager to co-ordinate the technical support. However, the further that scientists advance beyond the post-doctoral level, the more that their lives are devoted to doing paper work of one kind or another and managing the various activities of the laboratory. Very little time is spent at the bench. The more senior a scientist at the Hall Institute Institute is, the more likely that technicians and students in the laboratory will actually execute the experiments. It is fruitful to employ Knorr-Cetina's (1981) manufacturing metaphor and to think of a laboratory as consisting of blue-collar workers (technicians, mostly women) and white-collar workers (students, postdocs and senior scientists). Students and postdocs can be thought of as managerial trainees. Especially in American laboratories it is they who do the bulk of the bench work but at WEHI, it is the technicians.

Senior scientists are the managers of the factories which produce knowledge. So much so is this the case that at the WEHI, some worry lest scientists lose their experimental skills from lack of practice. Consequently, there is emphasis placed on the need for Ph.D. students to learn established techniques, usually from technicians and to do their own bench work. Still, to focus too closely on technical expertise is to miss the forest for the trees. The main skill of the working scientist is not in bench manipulation, but in experimental design, in research strategy and tactics. Again, this should not be seen as a inear skill. Successful scientists are like jugglers who have to keep their eyes on all the Indian clubs at once.

Clear beginnings and ends to research are uncommon. Experimental results are not neatly co-extensive with either papers or grants. Areas of interest normally develop and subside gradually. Over weeks and months, some ideas will peter out or be put on the back burner, some projects will suddenly burst into life with an unexpected advance. Experiments and stages within them, range dramatically in size, from an hour or two, conception to completion, up to days and weeks. Some can be more easily interrupted and resumed than others. Especially for the senior scientist who must co-ordinate and consult with technicians, students and post-doctoral fellows, no day is likely to pass with work focussed on one project only.

While strategic decisions involve what kinds of experiment to conduct, tactical decisions turn on who should do what next. A set of manipulable systems enables a number of experiments to go on concurrently. As we will demonstrate in detail below, these interactive pieces of research require further tactical decisions about which particular elements should be isolated and designated with the title 'experiment' for the purposes of presentation in a journal article. It is this rhetorical device, 'the experiment' which conceals the complex set of interactions between instruments, skills, techniques, raw materials, theory and social relations elements that we have called a manipulable system which together provide for the possibility of experimental data generation.

9. MANIPULATING THYMOCYTES

To exemplify many of the points made in the foregoing section, we look in some detail at one manipulable system within the LDU. In common with many other biomedical research organizations, WEHI provides in its *Annual Review* a list of its laboratories' concurrent projects. Though sometimes the distinctions are still rather artificial, these tend to represent the way the scientists themselves categorize their work and so, in that sense, are a natural taxonomy. Each project usually constitutes what we call a manipulable system. In mid-1987 there were eight in the LDU (WEHIMR, 1987: 112-6). The one on which we will concentrate is Scollay's main pre-occupation. The others involve collaborations between Scollay, his Unit Head, Ken Shortman, Martin Pearse (the Unit's molecular biologist), Ann Wilson (in charge of the Unit's support staff), together with workers from two other Units, two foreign and an Australian institution, visiting scientists, postdoctoral fellows and postgraduates. In the report of the work we are about to describe (Pearse *et al.*, 1989), Shortman, Pearse, Wilson and two postgraduates are listed as co-authors along with Scollay. The important roles of Scollay and his main collaborator, Pearse, are indicated by their listing as first and last-named authors respectively. In discussing Scollay's manipulable system, it needs to be born in mind that he is involved to some degree simultaneously in all but two of the LDU's projects.

Scollay's manipulable system involves studying T-cells at an early stage of maturation in the thymus; in particular, two groups of T-cells which lack the antigens CD4 and CD8 but certain sub-populations of which will subsequently develop into thymocytes which do have the CD4 and CD8 antigens. This process is tracked by injecting precursor T-cells into mice which have been irradiated to destroy their own T-cells. Periodically thereafter the mice thymuses are removed, the T-cells extracted and the stage of their maturation identified by which fluorescent marker antibodies bind to corresponding antigens. Identification is made through FACS machine data. Working in close collaboration, Pearse studies the gene re-arrangment associated with the differentiation using recombinant DNA techniques. The work began eighteen months before we interviewed Scollay. It took him a year to establish the system and he expected it to continue for a further year.

To establish the system, Scollay had learnt what the important variables were and how to manipulate them, to account for the effects given the age of donor and recipient mice, the amount of radiation needed to kill the recipient's T-cells, the number of cells to inject, when to assay, what to look for, how to identify and eliminate artefacts, and so on. Once the system was up and running (generating papers) its very productivity seized control of Scollay's research agenda. To get such a manipulable system working Scollay exploited one of WEHI's major strengths, its animal breeding, handling and maintenance facilities. A reliable supply of disease-free laboratory mice is often a hope rather than an expectation elsewhere. Irradiated mice, their immune system destroyed, pose a special problem. Scollay's own skill at making injections into the tiny mouse thymus, together with supervision of the FACS runs, are his major hands-on involvement. Most of the benchwork is done by his technician, Angela D'Amico. Scollay's most crucial role is the co-ordination and planning: ordering experiments, analysing the results and ordering fresh experiments.

Pearse *et al.* (1989) reports results generated by Scollay's system, discussing the results of experiments labeled 'A', 'B' and 'C', each of which is 'representative of two to four similar experiments.' We look at the history behind the two lines summarizing experiment 'C' to demonstrate the role played by experiment within a manipulable system. In experiment 'C' two groups of mice were injected with two sub-populations of T-cell precursors, each having a common surface marker, each lacking a common surface marker, one having and one lacking a third antigen. Experiments 'A' and 'B' were distinguished from each other and from experiment 'C' by the presence or absence of these same surface markers. Some of the mice were killed after seven days, the rest after eleven and their thymocytes checked for the presence or absence and frequency of CD4 and CD8 antigens. Scollay was trying to discover which antigen is significant in later CD4 and CD8 expression.

A one-page, handwritten note to D'Amico, ordering the experiment, indicates that the point is to determine the significance of the Pgp-1 antigen and that the experiment is a repeat of an earlier one. Scollay notes that the earlier experiment is numbered '100588'; D'Amico, that this one is 'exp. 30688'. These numbers relate to actual runs of Scollay's manipulable system, the appellations 'A', 'B' and 'C' having been generated for the purposes of the resultant paper which includes only 'significant' examples. D'Amico records her work in two separate places - an informal 'lab book' (a daily diary) and a formal 'experiment book' which records the details of experimental set-up and the subsequent assays. Experiment '30688' ('C') is recorded in six handwritten pages in D'Amico's experiment book. Three pages of the six give set-up details based on Scollay's order; the other three are based the record made in her diary of the assays. Under Scollay's supervision, samples of the thymocytes are run through the FACS machine both before they are injected and allowed to differentiate and after they have been recovered on the seventh and eleventh days.

The hardcopy FACS output is simply a 'dump' of what can be seen on the CRT display. Typically, it consists of four histograms showing the relative frequency of various cell sizes, together with the level of red, green and blue fluorescence. Beside this is a two-dimensional 'dot plot' of two of these four variables (e.g., cell size and green fluorescence), together with the relevant numbers [See Figures 1 & 2]. In this case, as

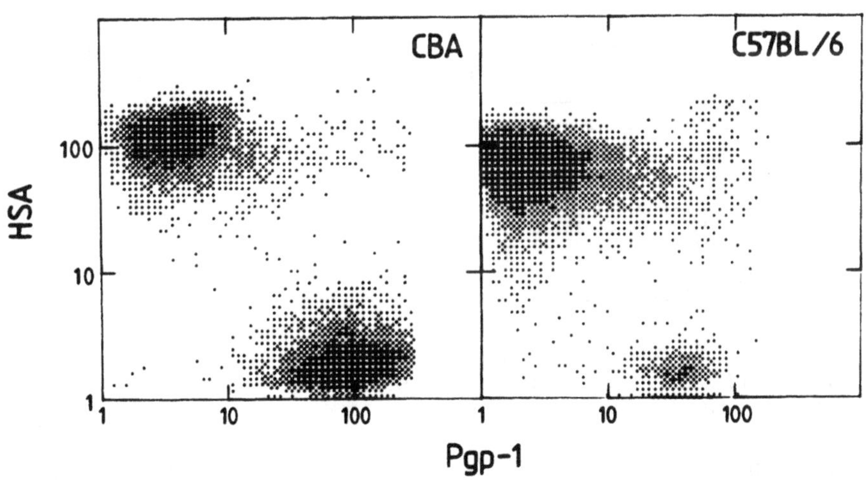

FIGURE 1: Two-colour immunofluorescence analysis of the expression of HSA and Pgp-1 antigens on the surface of CD4$^-$ CD8$^-$ thymocytes, showing two main subpopulations [HSA$^+$ Pgp-1$^-$ and HSA$^-$ Pgp1$^+$] with a trace of a third [HSA$^+$ Pgp-1$^+$]. Source: WEHIMR (1988: 91).

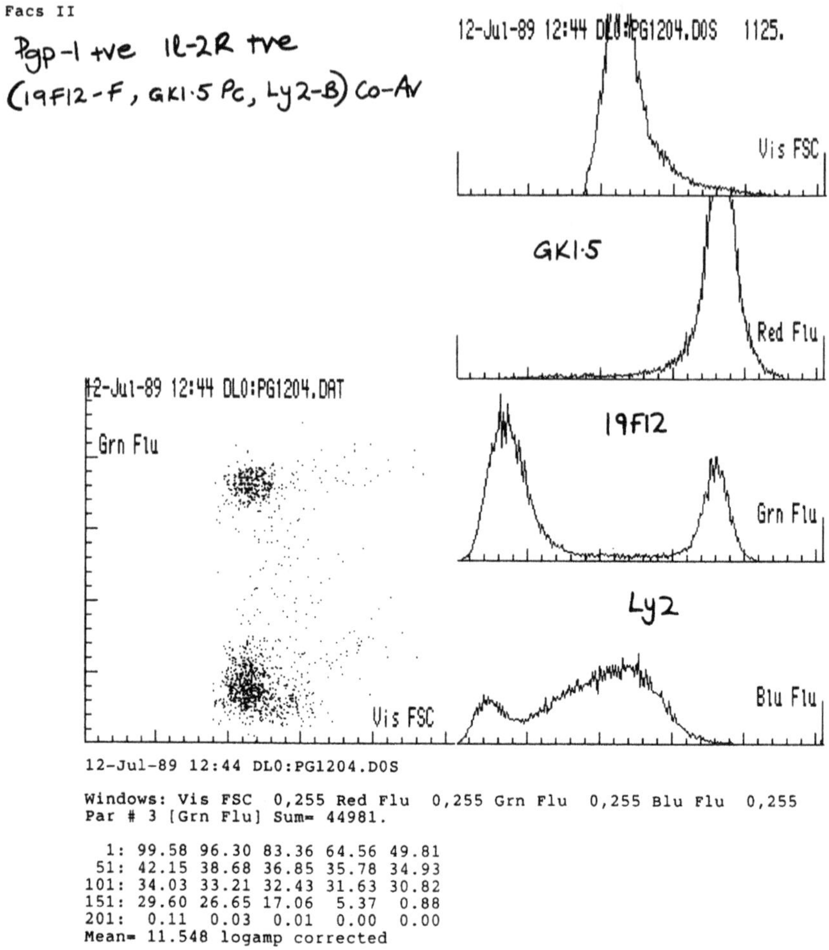

FIGURE 2: An example of FACS output, showing the relative frequency of fluorescent cells and a 'dot plot' of two of the variables, together with the numerical data. Source: Roland Scollay.

usual, Scollay re-analysed the raw data to compare various pairs of variables, both at 'set-up' and after days seven and eleven, generating further FACS printouts. These are annotated by Scollay, who also wrote summaries of the day seven and day eleven analysed assay data, as well as a summary of the experiment as a whole.

The two lines devoted to experiment 'C' in Pearse *et al.* (1989) thus summarize twelve pages of handwriting and five FACS printouts.

Experiment '30688' ('C') is certainly a distinguishable entity with firm boundaries: one can tell what data, what representations of that data and what analysis belong to it. It is an important experiment; after all, '30688' ('C') is, unlike countless others, reported in a subsequent publication . Even so, it is only 'representative of two to four similar experiments.' The main point to emerge from this discussion is that experiments may be usefully regarded as instances of the manipulation of a system, some of which - for one reason or another - turn out to be more important or interesting than others. In their portrayal of experiments, orthodox accounts of science tend to rely on published evidence, especially the article reporting original research results, in building up their picture of the scientific enterprise. But, as Medawar (1963: 378) pointed out and as the example we have examined in detail shows, the conventions governing scientific publication 'give a totally misleading narrative of the processes of thought that go into the making of scientific discoveries.'

10. THE MATERIALITY OF MANIPULABLE SYSTEMS

We have sought to emphasize that the products of manipulable systems are not only social constructs, but are the combined effects of human labour applied to materials through the application of tools and techniques. The data produced by this labour process is, as a consequence, materially constrained by reality: not in a way which reveals any necessary or certain truth about that reality, but in a way that reflects the condi ions of its production. In the cases we have examined this consists in getting the manipulable system to work. We concur with Knorr-Cetina (1981), Latour and Woolgar (1979) that much of what goes on in the laboratory is a judgement impregnated, local and contingent process of opportunistic tinkering. Science, in the words of the head of the Immunoparasitology Laboratory Graham Mitchell, is 'the art of the possible, studying what you can.' The material efficacy *and* contingency of science cannot be accounted for if the focus of analysis is limited to the interpretative flexibility of inscriptions. Manipulable systems are designed and operated to suit particular needs, but they must still be deployed to make experimental *enquiries* - the outcome of which he cannot be certain. Only in the poorer kind of scientific pedagogy does laboratory work have a guaranteed outcome.

We now need to expand the focus to recognize what goes on as a labour process, as the socially mediated but not socially determined generation of significant data. Further research needs to concentrate on the material conditions of data production: specifically, on the kinds of instruments and raw materials and the ways in which their use in manipulable systems produces specifically constrained kinds of data.

ACKNOWLEDGEMENTS

Quotations attributed to those at the Hall Institute, not otherwise sourced, derive from field notes and transcripts of interviews by the authors, as well as M. Charlesworth, L. Farrall and J. Reid (see Charlesworth *et al.*, 1989). We wish to acknowledge the open and extended co-operation the Director, scientific and support staff of the Hall Institute. In particular, we would like to thank A. D'Amico, K. Forsyth, R. Coppel, D. Kemp, G. Mitchell, R. Scollay, K. Shortman and A. Wilson. We are grateful to Mitchell and Scollay for reading and providing valuable comments on an earlier draft. It should not, however, be assumed that all of those whose work is discussed necessarily concur with what is said about it. The Financial support of the Australian Research Grants Scheme, Deakin University and the University of Wollongong is gratefully acknowledged, as is the advice and assistance of the Editor.

REFERENCES

Ackermann, R.J. (1985). *Data, instruments, and theory; a dialectical approach to the understanding of science*. Princeton: Princeton University Press.

Agassi, E. (1988). Do experiments depend on theories or theories on experiments? In Batens & Van Bendegem (1988), pp. 3-13.

Barnes, B. (1974). *Scientific knowledge and sociological theory*. London: Routledge.

Batens, D. & Van Bendegem, J.P., eds. (1988). *Theory and experiment: recent insights and new perspectives on their relation*. Dordrecht: Reidel.

Bhaskar, R. (1983). Experiment. In *Dictionary of the history of science*, ed. W.F. Bynum, E.J. Browne & R. Porter, pp. 136-8. London: Macmillan.

Bloor, D. (1976). *Knowledge and social imagery*. London: Routledge and Kegan Paul.

Burnet, F.M. (1968). *Changing Patterns, an atypical autobiography*. Melbourne: Heinemann.

Burnet, F.M. (1971). *Walter and Eliza Hall Institute, 1915-1965*. Melbourne: Melbourne University Press.

Charlesworth, M.J., Farrall, L.A., Stokes, T.D. & Turnbull, D. (1989). *Life among the scientists: an anthropological study of an Australian scientific community*. Melbourne: Oxford University Press.

Collins, H.M. (1985). *Changing order: replication and induction in scientific practice*. Beverly Hills: Sage.

Duhem, P. (1954). *The aim and structure of physical theory*. Princeton: Princeton University Press.

Galison, P. (1987). *How experiments end*. Chicago: University of Chicago Press.

Gilbert, G.N. & Mulkay, M.J. (1984). *Opening Pandora's box: a sociological analysis of scientists' discourse*. Cambridge: Cambridge University Press.

Hacking, I. (1983). *Representing and intervening*. Cambridge: Cambridge University Press.
Hughes, T.P. (1985). Edison and Electric Light. In *The social shaping of technology*, ed. D. MacKenzie & J. Wacjman, pp. 39-52. Milton Keynes: Open University Press.
Knorr-Cetina, K.D. (1981). *The manufacture of knowledge: an essay on the constructivist and contextual nature of science*. Oxford: Pergamon Press.
Knorr-Cetina, K.D. & Mulkay, M.J., eds. (1983). *Science observed: perspectives on the social study of science*. Beverly Hills: Sage.
Latour, B. (1986). Visualization and cognition; thinking with eyes and hands. *Knowledge and society*, **6**, 1-40.
Latour, B. & Woolgar, S. (1979). *Laboratory life: the social construction of scientific facts*. Beverly Hills: Sage; 2nd ed., 1986, Princeton: Princeton University Press.
Law, J. (1987). On the social explanation of technical change: the case of the Portuguese maritime expansion. *Technology and Culture*, **28**, 227-252.
MacKenzie, D. (1981). Some notes on the science and social relations debate. *Capital and class*, **14**, 47-60.
Medawar, P.B. (1963). Is the scientific paper a fraud? *The listener*, **70**, 377-8.
Mulkay, M.J. (1979a). *Science and the sociology of knowledge*. London: Allen & Unwin.
Mulkay, M.J. (1979b). Knowledge and utility: implications for the sociology of knowledge. *Social studies of science*, **9**, 63-80.
Mulkay, M.J. (1985). *The word and the world: explorations of the form of sociological analysis*. London: Allen & Unwin.
Mulkay, M.J., Potter, J. & Yearley, S. (1983). Why an analysis of scientific discourse is needed. In Knorr-Cetina & Mulkay (1983), pp. 171-203.
Pearse, M., Li, W., Egerton, M., Wilson, A., Shortman, K. & Scollay, R. (1989). An early thymocyte developmental sequence marked by transient expression of the IL-2 receptor. *Proceedings of the National Academy of Science (U.S.A.)*, **86**, 1614-18.
Pickering, A. (1985). Pragmatic realism and the macro-sociology of experiment. *Abstracts, BSHS/BSA Conference on 'The uses of experiment'*, Bath.
Pinch, T.J. (1986). *Confronting nature*. Dordrecht: Reidel.
Reichenbach, H. (1951). *The rise of scientific philosophy*. Berkeley: University of California Press.
Rouse, J. (1987). *Knowledge and power; towards a political philosophy of science*. Ithaca: Cornell University Press.
Shapin, S. & Schaffer, S. (1985). *Leviathan and the air pump; Hobbes, Boyle and the experimental life*. Princeton: Princeton University Press.
Studer, K.E. & Chubin, D.E. (1980). *The cancer mission: social contexts of biomedical research*. Beverly Hills: Sage.
Trager, W.T. & Jensen, J.B. (1976). Human malaria parasites in continuous culture. *Science*, **193**, 673-5.
Turnbull, D. (1984). Relativism, reflexivity and the sociology of scientific knowledge. *Metascience*, **1/2**, 47-60.
Turnbull D. (1989). The Push for a Malaria Vaccine. *Social studies of science*, **19**, 283-300.
WEHIMR. *The Walter and Eliza Hall Institute of Medical Research Annual Review*. Melbourne.
Westmore, A. (1983). Gene splicing brings hope of malaria vaccine. *New scientist*, 760.
The Wistar Institute of Anatomy and Biology (1984) annual report. Philadelphia.
Young, B. (1979). Science is a labour process. *Science for people*, **43/44**, 31-37.

Young, B. (1985). Is nature a labour process? In *Science technology and the labour process*, ed. L. Levidow & B. Young, *Marxist studies*, 2, pp. 206-31. London: Free Association Books.

Deakin University
University of Wollongong

STEPHEN GAUKROGER

EXPERIMENT AND THE MOLECULARITY OF MEANING

The account we are able to give of the role of experiment in scientific theories depends to a large extent upon the account we offer of the meaning of scientific statements. In what follows, I consider four theories of meaning.[1] These can be represented as follows:

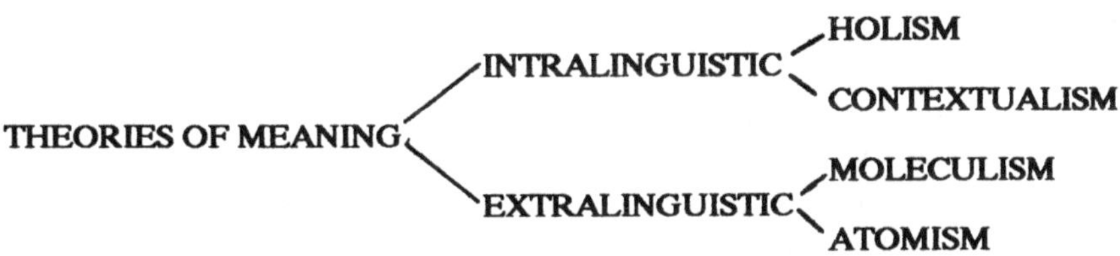

On intralinguistic accounts of meaning, meaning consists in a relation between linguistic items. Reference plays no essential role in accounting for meaning. On extralinguistic accounts, meaning consists in a relation between linguistic and non-linguistic items and reference plays an essential role in accounting for meaning. I identify two types of intralinguistic and two types of extralinguistic theory. The most straightforward way of distinguishing these is in terms of what they pick out as the basic or minimal unit of meaning or empirical significance. For atomism this is the word or term; for moleculism it is the sentence or individual theory; for contextualism it is the whole body of theory (or specific language or conceptual scheme); and for holism it is the whole of science (or language or thought).

In disputes over meaning in the history and philosophy of science, holism and especially moleculism have been largely ignored. Contextualism and atomism have been pitted against one another, the former being taken as constitutive of the intralinguistic approach; the latter, of the extralinguistic approach. Since the work of Koyre, and more recently Kuhn, this has worked very much in favour of the former. Atomism, because of its attempt to associate terms directly with objects, events or states of affairs, could provide little basis for exploring the systematic structural features of large-scale bodies of theory. Moreover,

H. E. Le Grand (ed.), Experimental Inquiries, 193–213.

this attempt carried with it a commitment to a conception of scientific development as being piecemeal and cumulative, something which was completely at odds with the existence of such major and apparently discontinuous changes in the development of science as scientific revolutions. Contextualism, on the other hand, at least as it was taken up in the philosophy of science, was virtually designed around the idea of a scientific revolution and Koyre and others took it as their principal aim to provide a plausible account of the discontinuous development of science.

The success of contextualism was twofold: that approach not only offered a more plausible account of scientific structure and development, but it also made those issues *the* issues in the philosophy of science. Because it managed to do both these things, the phenomenon of 'spoils to the victor' occurred. The contextualist account of a range of other issues was accepted as part of the package. Particularly affected here was the question of method: the inductivist programme, which made experiment an ingredient in scientific discovery, was rejected in favour of approaches where one begins with hypotheses or conjectures, and where experiment enters the picture, if at all, only in testing such hypotheses or conjectures.

This approach was enshrined in the distinction between 'the context of discovery' and 'the context of justification', the general thrust of which was that discovery is a conceptual or intellectual process whereas justification is an empirical one. Underlying this distinction, I believe, is a semantic assumption that what it is that gives an empirical statement meaning is different from and prior to what it is that makes, or would make, that statement true. Experiment and observation play no role in establishing meaning, but they may play a role in establishing whether meaningful empirical statements are true. Atomists have occasionally rejected this assumption. But, their attempts to provide something better have failed. Operationalism, for example, which was one of the few concerted attempts to account for the close ties between experiment and the formulation of theories, ended up imposing the completely unrealistic and indeed impossible requirement that every theoretical term be associated with an overt experimental procedure. The failure of this approach only served to reinforce the *prima facie* case for contextualism.

In the history of science, the contextualist approach is best illustrated by the extreme form it takes in Koyre's work. Koyre construed the seventeenth-century scientific revolution in physics and astronomy in terms of a conflict between a mathematically-orientated Platonism and an experimentally-orientated Aristotelism, where the former completely triumphed. In particular, he put forward the now famous argument that the role of observation and experiment in Galileo's astronomical and mechanical theories was minimal, that he actually performed few of the experiments he claimed to have performed, and that in general his procedure was the Platonistically-inspired one of subordinating empirical reality to a realm of mathematical idealizations (Koyre, 1968; 1978). Galileo's account of falling bodies, for example, which is the cornerstone of his kinematics, is formulated in terms of a law that describes a situation -- bodies falling in a void -- which no-one has ever experienced and which never occurs. Koyre presents it as an idealization which abstracts from real bodies so as to exhibit a simple mathematical relation between time and distance traversed. Experiment has no role to play in this process, as he construes it. Moreover, this idea that Galileo's mathematical physics is possible only through mathematical idealization provides a good deal of implicit support for Koyre's contextualism and may indeed motivate it to some extent. For while an atomist conception may seem plausible, and perhaps even natural, in the case of an experientially-based physics, Koyre's argument is that such a physics is alien to Galileo, and more generally to the whole of modern science. Galilean physics is the physics of a world of mathematical idealizations, as indeed is all mathematical physics, on Koyre's view, and atomism is not especially plausible or natural as a theory of meaning for mathematical statements. Some form of relational element enters here, as indeed it seems to do when we consider fundamental quantitative concepts like 'mass', 'force', 'acceleration', *etc.* In other words, in providing an account of the meanings of the statements in a physical theory, a good deal hinges on whether that theory is one which is expressed in mathematical terms and, most importantly of all, what one thinks the conditions of possibility of such mathematical expression are. For Koyre, the mathematization of physics is effectively a translation of physics into the Platonic realm of numbers and geometrical figures. Hence it is not too surprising that a conception of meaning with no strong appeal in the case of physics *simpliciter* but with considerable appeal in the case of mathematics should be advocated. And this conception is one on which the meaning of a term is a function of its place within a complex network of interrelated terms, and has nothing to do with questions of reference.

In what follows, I want to show three things. The first is that experiment may on occasion play an indispensable role in the discovery and formulation of theories. Second, in so far as we want a theory of meaning to help us elucidate this role, we must first decide which theories of meaning are viable. I will show that contextualism and atomism are not viable as theories of meaning, the first because it is internally incoherent, the second because it cannot provide a proper basis for a semantic theory. Consequently, I will argue, if there is a real issue at stake between intralinguistic and extralinguistic conceptions this issue is best seen in terms of holism and moleculism. Third, a moleculist theory of meaning can be used to provide us with a detailed understanding of the way in which experiment functions in the elaboration of new theories, whereas on a holist account it is virtually unintelligible how experiment could play this role.

1. EXPERIMENT AND THE FORMULATION OF THEORIES: THE CASE OF FALLING BODIES

To illustrate the role of experiment in the formulation of theories I shall describe one case in some detail: Galileo's formulation of the law of falling bodies.[2] This is a particularly good example because it enables us to see clearly, without any 'rational reconstruction' on our part, how experiment plays a role in the discovery of the law that all bodies fall at the same rate of acceleration in a void.

In *Two New Sciences*, Galileo provides the first modern full-scale kinematical treatment of motion: in particular, he presents and justifies laws concerning free fall and projectile motion. These he presents in the form of mathematical descriptions of what happens in a void. Now the motion of bodies in a void is something we never experience and something to which we have no direct access. The motions of bodies in resisting media *is* something we regularly experience, yet these motions differ from the motions those same bodies would undergo in a void. Galileo's law of free fall tell us that all bodies undergo a uniform acceleration in a void but this is clearly not the case in a resisting medium. At first sight, therefore, the law appears to suffer from two drawbacks: it appears to tell us something about a situation which may never occur, and it appears not to tell us about situations which do normally occur. Hence there seem to be problems both about the relevance of the law and about whether it could receive any evidential support. This type of problem affects many general laws, and it has been the source of many philosophical disputes. Since we do not experience bodies falling in a void, for example, we cannot arrive at Galileo's law inductively. On the other hand, since, if the law holds, its truth is contingent -- bodies may just as easily have fallen at different rates in a void, or may have fallen with an unaccelerated or nonuniformly accelerated motion -- it is impossible that *a priori* arguments will lead us to the law. To maintain that the law is an hypothesis open to empirical tests is of no real help either. First, the problem simply reappears at a different level. The situation described in the law does not naturally occur and cannot be experimentally induced, so in what way is it open to empirical test? And, even if the situation *could* somehow be experimentally induced, that would still leave the problem of how the law could be at all relevant to the case of bodies falling in resisting media. Second, the presumption behind the hypothetical construal is that the theory itself is somehow developed at a purely conceptual level and then tested empirically to determine whether it is true or not. But, experiment is an integral part of scientific discovery, not an added extra.

This is, at least, the view presented and defended in Galileo's own account of how the law is to be established (1974: 65-108). Galileo's project is to discover the relations that hold between a body falling in a void and that body falling in a resisting medium. He takes the fall of bodies in a resisting medium as his starting point and then describes a series of experiments, including thought-experiments, designed to decide what factors are operative in determining the rate of fall of a body and how these factors operate.

He deals first with the traditional Aristotelian view that rate of fall is directly proportional to absolute weight. He has two arguments against this. The first is empirical: if two bodies made of the same material but of different absolute weights are dropped simultaneously from the top of a tower to the ground, they arrive at the ground simultaneously. The second is a thought experiment. If we drop, say, two lead spheres of different weights, then on the Aristotelian account the heavier will fall faster. But, suppose we tie the spheres together. The slower one would then surely slow down the faster one, and the faster one would speed up the slower, so that the resultant speed would be somewhere between the two original speeds. But the aggregate weight is greater than the weight of the heavier body. Hence, rate of fall cannot be directly proportional to absolute weight. Now Aristotle had also held that the rate of fall is inversely proportional to the density of the medium. Against this another thought experiment is proposed. If we let the density ratio of water to air be n:1, where $n>1$ (since the specific weight of water is greater than that of air), and take a body which falls in air but floats in water (*e.g.*, a wooden sphere), and say that this has a rate of fall of one unit in air, then it would follow that it has a rate of fall of $1/n$ units in water. But, we have already said that it floats -- *i.e.* would rise and not fall, in water. So rate of fall cannot be inversely proportional to the density of the medium.

Next, Galileo makes an important generalization. Instead of thinking simply in terms of rate of fall determined with respect to one body in two media or with respect to two bodies in the one medium, he considers the case of any body in any medium. First, he describes an experiment which shows that the ratio between the rates of fall of bodies is not the same as the ratio of their specific weights. Gold and lead, which fall at approximately the same rate in air, behave quite differently in mercury, the former sinking, the latter rising to the surface. This experiment indicates that differences in rate of fall of bodies diminishes as the density of the medium decreases. This prompts him to ask what would happen in the limiting case of a void: he raises the possibility that in such a case the rate of fall of all bodies would be equal. But, until we know the precise connection between speed, specific weight and resistance, we will not be able to establish this. Galileo therefore proposes an experiment in which the buoyancy effect of the medium can be distinguished. The buoyancy effect is the ratio between the specific weight of the body and that of the medium. The problem is to determine precisely what effect this ratio has on rate of fall. He compares the buoyancy effect of two media (air and water) on two bodies (ebony and lead). Given the specific weights of these substances, the buoyancy effect can easily be calculated. It turns out that the buoyancy effect varies much more radically than the specific weight of the body: if we let the specific weights of air, water, ebony and lead, be 1, 800, 1000 and 10000 respectively, then it turns out that whereas the buoyancy effect of air has a negligible effect on rates of fall of ebony and lead, the buoyancy effect of water on ebony is huge (it loses four-fifths of its effective weight) whereas its effect on lead is very small (less than one-tenth). It is the specific weight *ratios* that determine rate of fall, not the specific weights themselves. Since a void has no specific weight it cannot bear a ratio relation to the specific weight of the falling body; *i.e.*, this ratio which determines differences in the rate of fall cannot be operative in the case of a void and so we must conclude that all bodies -- whatever their specific weight -- fall in a void with the same 'degree of speed' (*i.e.*, as it turns out, degree of uniformly accelerated motion). This conclusion is particularly important since on the basis of the equality of rates of fall of all bodies in a void we can proceed, at least in principle, to calculate differences in speeds between any two bodies in any media by determining the amount by which the theoretical speed in a void will be diminished by comparing the specific weight of bodies with that of the media. To this end, Galileo takes us through (largely unsuccessful) experiments to measure the specific weight of air.

There remains one problem. Bodies falling in media do not in fact accelerate uniformly. Neither specific weight nor the buoyancy effect can account for this since they are both constant (the latter being a constant for any particular body and any particular medium). This leads Galileo to invoke a form of resistance to fall which is distinct from the buoyancy effect: the friction effect. The friction effect increases with the acceleration of the body since larger and larger amounts of resisting medium have to be traversed, and equilibrium is reached when the body ceases to accelerate because of friction. This state of equilibrium occurs much earlier in rarer bodies, not because the friction effect bears a direct relation to specific weight, but because the buoyancy effect is much greater in bodies of lower density and hence their motion is already greatly retarded. For this conclusion to go through, however, two things have to be shown: first, that the friction effect is greater for rarer bodies; second, that it increases with speed. In order to show the first we need to isolate the friction effect experimentally from the buoyancy effect. Free fall does not allow us to do this. Galileo suggests rolling two bodies, *e.g.* one of cork and one of lead, down a plane which is gently inclined so as to make the motions as slow as possible and thereby to reduce the buoyancy effect. The trouble here is that the more gentle the incline the greater the surface friction, which would interfere with our isolating the friction effect (which is totally different from surface friction since it is an effect of the medium). He resolves these difficulties by proposing an ingenious experiment in which a cork and a lead sphere are suspended on threads of equal length and set in oscillation. The periods of oscillation remain identical for both spheres but in the case of the cork sphere the amplitude of swing is considerably reduced very quickly. This cannot be due to the greater specific weight of the lead causing it to move faster: we can begin the experiment by swinging the cork through a greater arc so that it initially moves faster but the same thing will happen. Moreover, since the buoyancy effect is simply the specific weight ratio, it cannot be due to this either. It must therefore be due to the friction effect. Finally, all that remains to be shown is that the friction effect increases with speed. Again no direct experiment on freely falling bodies is possible because of the great distances that would be involved and the difficulties in measurement that would ensue. Hence the consequence that bodies projected at artificially high speeds will be retarded until they reach their natural maximum speed for that medium is of crucial importance, since an experimental situation in which this can be tested can be realized in a relatively easy and straightforward way. We simply fire a gun vertically downwards from a great height and measure the penetration of the bullet into the ground. We then fire the gun close to the ground and measure that penetration. The first is less than the second, which means that the bullet has been retarded.

In sum, Galileo shows, by means of a series of actual experiments and thought experiments, that rate of fall bears a complex relation to specific weight, buoyancy effect and friction effect. By determining exactly how these factors are related to one another he is able to determine what happens when the medium is removed entirely, and this forms the content of his law of fall. This is, I maintain, a clear example of a case in which experiment plays a necessary role in the formulation of a theory, in this case the law of free fall. Having established the existence of such cases, what we need to provide is an account of what precisely the role of experiment is in these cases. Our ability to give such an account is, however, dependent on our having the right theory of meaning, as I shall show, so first we need to decide between the claims of the various theories of meaning that I have outlined.

2. THEORIES OF MEANING

Let us begin with intralinguistic theories. Contextualism can be dispensed with quickly, as Davidson (1984: ch. 13) has already identified a fatal flaw in contextualism as a theory of meaning. The problem derives from the idea that meanings are determined wholly within specific theoretical contexts or conceptual schemes and that they cannot transcend particular contexts or conceptual schemes. If we are asked what makes different conceptual schemes different, we cannot simply answer that they operate with different concepts. For the question then arises of how we tell that the concepts are genuinely different: what assures us that they are not merely notational variations of the same concepts? To be able to distinguish the conceptual schemes from one another we need to be able to compare them with something else in respect of which they differ from one another. The most straightforward candidate for this role is their subject-matter. Let us say that we are dealing with physical theories and that their subject-matter is physical reality. The problem is that, on the contextualist account, we have no independent access to this subject matter: physical reality is conceived as something undifferentiated in its own right, something which is carved up differently by different conceptual schemes. On a relativist version of this view, it will be argued that we have no way of choosing between what are effectively different 'mappings', whereas on a non-relativist version it will be argued that we have a way of choosing, albeit an internal one, because some mapping will yield better predictions, will solve a greater range of problems, or whatever. But in fact neither of these claims is sustainable. As I have said, we cannot maintain that reality is carved up or mapped out differently (or in the same way) by conceptual schemes unless we have something by which to distinguish those schemes. On the contextualist account, this is precisely what we cannot have, for we have no access to reality other than via the conceptual scheme: if we had, we would not need the conceptual scheme, for we would be able to go directly to reality. But, if we have no independent access, we cannot say that it is the one reality that the conceptual schemes carve up differently (or in the same way). Now suppose that, in response to this, one were to maintain that each conceptual scheme creates its own subject matter, or even that it creates its own reality. This fares no better. For again we cannot say what makes these subject matters or realities different (or the same). Without criteria for 'same reality' or 'same subject matter' or for 'different reality' or 'different subject matter', we are unable intelligibly to articulate the contextualist view. There is something fundamentally flawed with the language of conceptual schemes, and thereby with contextualism. By the very nature of the case, we cannot have criteria for distinguishing conceptual schemes: no purely internal criteria will succeed, and external ones are ruled out because we only ever have access to anything other than the conceptual schemes by means of those conceptual schemes. And, where we have no criterion for distinguishing them, we cannot talk about different conceptual schemes.

Note that this failure of contextualism is a failure *as a theory of meaning*. This does not mean that the bodies of theory that contextualism picks out as forming coherent wholes do not form coherent wholes, only that if they do form coherent wholes this cannot be on the basis of some supposed identity provided by an exclusive network of interrelated meanings. But, they may have a collective identity on other grounds. After all, even semantic atomists have to distinguish large bodies of theory -- say Aristotelian physics and Newtonian physics -- from one another. An atomist theory of meaning does not allow one to construe this collective identity in terms of belonging to the same network of meanings, of course, but this does not mean that the atomist has no basis on which to pick out large bodies of theory: shared subject-matter, appeal to a shared body of evidence, shared commitment to particular methodologies, all serve to separate and unify individual theories.

In response to the failure of contextualism there are two directions in which one can go: to opt for holism or for an extralinguistic account. The former anchors meanings in one true grand theory; the latter, in reference. Davidson opts for holism. He offers a truth-conditional account of meaning but his notion of truth does not invoke the idea of reference, and considerations of meaning are independent of and prior to considerations of reference. He rejects theories which try to account for 'true-in-L', and his own theory of meaning takes 'the form of a theory of absolute truth' (1984: 225). It cannot be denied that, at the phenomenological level, there exist different theories applying different concepts to the same subject matter, but on Davidson's view they all essentially aim to achieve the same thing despite appearances. One is to be guided in this respect, Davidson argues, by his 'Principle of Charity' which dictates that interpretation should always aim to maximize agreement between our own theory and those we are interpreting. Davidson is understandably coy about to what this agreement may amount (1984: xvii) but the thrust of the doctrine seems to be that we are to interpret all theories in terms of our own best theories and to interpret them in such a way as to maximize the number of true statements -- these being those statements that our own theory tells us are true -- that they make.

On the face of it, this is a radically Whiggish claim. Its force is weakened, however, by the fact that the notion of 'meaning' employed in holism, at least in its Davidsonian form, is a very rarefied notion which has no connection with questions about how, for example, meaning and truth might be established. Consequently, differences arising from the way in which claims are advanced and defended in different theories, or from the way in which truth is supposedly to be established and tested for, fall outside Davidson's strictures (*cf.* Hacking, 1982; 1985). This makes holism somewhat peripheral to substantive issues in the history and philosophy of science but I believe it is going too far to maintain, as Hacking (1983) does, that questions of meaning generally are irrelevant to the substantive issues. The fact that holism is irrelevant in this context counts against it and the rarefied conception of meaning it employs, for, as I shall show, questions of meaning are absolutely indispensable to our understanding of at least one substantive issue, namely experiment, and for this the more substantial conception of meaning offered by moleculism is needed. Before we look at moleculism, however, let us consider the problems with atomism.

Atomism is the traditional form taken by extralinguistic accounts of meaning from Plato's *Cratylus* onwards and its model or prototype is naming. Frege retained the naming model but showed that an account of meaning rich enough to capture those syntactical aspects of language on which inference depends would have to take the sentence, and not the term or the word, as its minimal unit of meaning. It is important that we understand in this respect what exactly the problem with atomism is. The difficulties arise most clearly in the context of expressions of generality. Classical syllogistic was restricted in its treatment of generality to sentences containing one expression of generality (*e.g.*, '*Everybody* is mortal'), but attempts were made by Medieval logicians to provide a more powerful system of syllogistic logic, one that could cope with sentences involving multiple generality (*e.g.*, '*Everybody* loves *somebody*'). Such expressions are extremely difficult to deal with syllogistically, and Medieval logicians hit upon the idea of construing expressions of generality as affecting, not the terms themselves or what they stand for, but their 'supposition', *i.e.* the way in which they signify what they stand for. The difference between the expression 'all men' and 'some men', for example, is analyzed in terms of 'all' and 'some' modifying the way in which 'man' stands for the class of men, not in terms of a modification of what the term stands for, as was classically thought. But this kind of approach proved useless: no uniform treatment was possible for all cases because modes of supposition varied with the complexity of the sentence, owing to the fact that every species of 'nesting' (the inclusion of the scope of one expression of generality within the scope of another) had to be accounted for differently, and in what was in effect an *ad hoc* way.

The problem of multiple generality is not an isolated failing of syllogistic theory, but an indication of a general problem with atomism. In syllogistic theory, sentences are constructed simultaneously out of their components. This simply mirrors the linear ordering of signs in sentences and obfuscates deep differences between such expressions as 'Oedipus envies Lafus', which consists of two proper names related by a verb, and 'Everybody envies somebody', which is treated in syllogistic theory on the model of the first, with expressions of generality functioning like proper names: traditional logicians were simply misled by superficial features of natural language. In response to this, Frege developed a notation which not only had no misleading superficial features of this kind but, more importantly, was able to capture the constructional history of sentences in their formal analysis. For example, the formal analysis of the sentence 'Everybody envies somebody' shows how it is constructed in several stages from sentences in which proper names are removed and the resulting predicates combined with expressions of generality. To this end, Frege replaced the notation used to express generality in natural languages with a new notation, that of quantifiers and variables. What this enables one to do is to use a straightforward account of the truth conditions of sentences with single generality as a universal account covering all types of generality by restricting the introduction of expressions of generality to the appropriate stages in the construction of the sentence. In applying this conception of the step-by-step construction of sentences to the problem of how generality is expressed, Frege invoked the notion of one-place predicate, thought of not as being synthesized from its components, but as formed by omission of a proper name from a sentence. The idea that sentences are synthesized from their components is one which had dominated the syllogistic tradition. On the syllogistic account, the complexity of sentences derives from the complexity of the subject and predicate concepts from which the sentence is composed. In rejecting this type of account, and developing a notion of sentential complexity adequate to capturing patterns of inference, Frege construes the complexity of a sentence as being due to an inner sentential complexity; *i.e.*, that the sentence in question is constructed out of a number of logically simple sentences, where a logically simple sentence is one which is not composed of subordinate expressions each possessing its own sense and contributing to the sense of the whole.

In sum, Frege's attempt to provide a semantics of language led him away from traditional atomist conceptions. This semantics is now the one generally accepted (allowing that problems remain in areas such as indirect speech and higher-order expressions). It is incompatible with atomism, but not with holism: a holist could allow that sentences are the minimal units of meaning but maintain that they have this meaning only in virtue of their relation to all the other sentences in language. This is not Frege's view: he construes the meanings of words holistically, in that the meaning of a word is a function of every true sentence in which that word figures, but not sentences themselves. As regards the minimal independent units of meaning, sentences, Frege holds to the naming model and hence to an extralinguistic conception of meaning.

Frege's account has two elements which, as Dummett (1981a: esp. chs. 6 & 7) has shown, must be kept distinct. The first is a theory of the different semantic roles of expressions. This theory is designed to tell us how the constituent expressions of a sentence determine that sentence as being true or false. It is the semantic values of the individual expressions making up the sentence that determine what the semantic value of the sentence will be; *i.e.*, given Frege's classical two-valued logic, whether the sentence will take the semantic value 'true' or the semantic value 'false'. The semantic value of each logical type of expression is determined in a different way depending on its logical type, the crucial question being what properties we have to ascribe to expressions of each type in order to determine the sentence as being true or false. These properties make up the semantic role of that type of expression. To say that an expression ha a particular semantic role is to say that it has something of the appropriate type associated with it -- objects in the case of proper names, functions from objects to the semantic values of sentences in the case of predicates, and so on for the other varieties of expression -- where the 'something of the appropriate type' associated with it we call its semantic value. Such an account provides us a way of assigning interpretations to formulae and thereby with a way of demonstrating properties of formulae and formal systems such as validity, soundness and completeness.

Now the semantic value of an expression is its reference for Frege, and since every type of constituent expression in a sentence contributes to the truth value of every sentence, every type of constituent expression has a semantic value; *i.e.*, has a reference. But Frege does not just conceive of reference in purely semantic terms as semantic value; he also conceives it in terms of the name/bearer prototype. This is the second distinct element of Frege's account, and it is this element that I want to focus on and develop, for this is where moleculism becomes a distinctive doctrine and one of relevance to the question of experiment. Once we equate reference with the notion of a bearer, it is no longer possible to restrict oneself simply to pointing out what types of reference the expressions of different categories have and then showing how these combine in a sentence to determine the truth value of the sentence. We now have to ask how we determine, in the prototypical case, what the reference of a proper name is. More generally, what does someone who can pick out the bearer of a proper name have to know in order to be able to do this? In what does that person's understanding of the name consist? Frege's answer is that it consists in a grasp of the sense of the name. The sense of an expression can be defined as constituting the manner in which something is determined as being its referent, so that in grasping the sense of an expression we understand the condition for something being its referent. In the case of proper names, for example, we can think of the name as being the route to its referent, provided we add the crucial qualification that this 'route' cannot be conceived in terms of a pre-given goal. Following Dummett (1981b: ch. 3), we can think of a name as determining the route, and its route as determining the terminus of that route, namely the referent. This leaves open the possibility that the route may in fact lead nowhere; *i.e.*, the proper name may be non-referring in that it has no bearer.

The 'route' that the sense of the proper name provides consists in criteria of identity for the bearer of the name. What is at issue here shows up clearly when we compare Frege's account with Mill's. For Mill, the use of proper names simply involves the bare association of proper names with their referents; the world is already differentiated and the point of the exercise is to label its components. For Frege, on the other hand, we are never simply given an object: it must always be presented to us in some way. Indeed, to say that something is an object is to say that we can refer to it by means of a proper name (or an expression which functions like a proper name) and it is the sense of the proper name that will determine what the bearer, if any, of that proper name is, by providing criteria of identity for the bearer. In the prototypical case where the bearer is a concrete object, two things are involved: our grasp of the linguistic significance of the proper name and our apprehension of an object which we can perceive or discern in some way. These two factors can easily be conflated under the general rubric of 'sense', but there are in fact two quite distinct modes of presentation involved. We should not run together the mode in which an object is presented in observation and the mode in which it can be said to be presented to us when we grasp the sense of its name. The former is required as something distinct because, amongst other things, we must have a means of picking out an object as the referent of a proper name for which no further question of identification arises. This means, in the prototypical case, ostension. This is not to say that ostension is infallible, that it requires no conceptual grasp or that it is 'perspectiveless' knowledge; only that where there is doubt about an ostension, *e.g.* doubt about whether an object has been picked out, this doubt is not to be resolved by changing to another mode of presentation, in particular by appeal to the criteria of identity supplied by the sense of a proper name. If the two modes of presentation were the same, or if disputes over ostension were to be resolved by appeal to another mode of presentation, we would never be able to leave the realm of sense and our theory would consequently not be extralinguistic but intralinguistic.

This 'double mode of presentation' conception, whereby an intralinguistic and an extralinguistic means are required for the identification of a referent, will form the core of my approach to the understanding of the role of experiment. Before we turn to experiment, however, it will be worth noting a few general features of the doctrine, including some of its more problematic aspects.

The first thing to note is that ostension is not the only way in which we can indicate or grasp the existence of a bearer other than by means of simply considering it as being the referent of an expression of a particular type. Numbers, for example, are abstract objects on a Fregean account. They obviously cannot be picked out ostensively, but there is a particular mode of presentation of numbers which is analogous to the ostensive mode of presentation of concrete objects, namely their presentation as numerals. When we say that some arithmetical function, *e.g.* x^2+2x,is computable for some given argument, *i.e.* for some given number, then, as Dummett points out (1981b: ch. 7), we mean the number to be given in the form of a numeral in binary notation, not in the form of 'the population of Hawaii'. Two general principles link this case with that of concrete objects. The first is that a criterion for identifying something as true concerning some object in a range of objects cannot genuinely relate to the objects in that range *simpliciter*, but must relate to the object considered as presented in some way. There is no such thing as judging something to be true of an object unless that object has first been singled out in some way. This principle marks out moleculism from atomism. The second principle marks out moleculism from contextualism and holism. It holds that for each category of referents there is some favoured mode of presentation. This is ostension ib the case of concrete objects and presentation in the form of numerals in the case of numbers. The favoured mode of presentation presents the object in such a way that no further question of identification can arise, or at least no further question of identification which requires us to invoke a different mode of presentation.

There are problems with the name/bearer prototype once one considers abstract objects such as shapes, and especially when one turns to incomplete expressions such as predicates. Moreover, in the mathematical case, the notion that one recognises the truth of a numerical equation by identifying an external object as the referent of a proper name would be extremely counterintuitive. It is such problems that have most often given rise to disputes over whether a realist theory of meaning is viable. I shall not consider these disputes here. My concern is, rather, with the case of concrete objects, but concrete objects as described in general laws and not identifiable ostensively. The 'double mode of presentation' doctrine may be thought to break down in these cases. As I now hope to show, however, in fact this is where the doctrine shows its mettle.

3. MEANING AND EMPIRICAL SIGNIFICANCE

If we wish to capture what Galileo's procedure achieves in terms of a theory of meaning, then the most natural approach is one utilizing the idea of double mode of presentation, something which is distinctive of moleculism. Consider the difference between this and a holist approach, for example. For the holist, one's grasp of the meaning of a sentence is a question which is independent of and prior to that of how we identify the referents (if any) of the sentence. At the level of theories of method, this is reflected in the view that hypotheses are formulated independently of and prior to questions of how we might determine their truth. On the moleculist account, on the other hand, there are two distinct factors in grasping the meaning of a sentence. First, we must have some way of grasping the sense of the sentence, which means that we must be able to grasp the conditions under which what that sentence states is true and the conditions under which it is false. Second, we must have some way, which is independent of a mere grasp of the sense of a sentence, of identifying to what it is that the sentence refers. In the case we are concerned with, where we are dealing with spatio-temporal objects, this means of identification will be empirical: ostensions in the prototypical case of directly accessible objects and situations, more complex procedures in other cases. Galileo's procedure is an example of the latter, and it enables Galileo both to formulate and legitimate (or, more strictly speaking, show one how to legitimate) the law in the one procedure. There are two conceptually distinct aspects to this procedure, but they are complementary.

What I have called the double mode of presentation approach conceives of the grasp of the meaning of a sentence in terms of getting a fix on the referent by means of two independent but complementary routes. It involves two ways of grasping referents, one purely conceptual and the other (in the kind of case we are considering) empirical. The first is independent of our perceptual abilities, instruments, experimental ingenuity, *etc.*, whereas the second is dependent upon those things. The reference is fixed, and the sense of the sentence made wholly determinate, when the two paths meet. They can be compared to paths from the orthogonal axes of a graph which uniquely determine a particular set of coordinates. Verificationists mistakenly ignored the first kind of grasp. Owing in large part to its dependence on an atomist conception of meaning whereby meaning consists in the bare association of proper names and referents, verificationism, in moving from the prototypical case of naming to the case of perceptually inaccessible states of affairs, advocates the idea that meaning is given by the mode of verification. But to ignore the second type of grasp, as intralinguistic theories do, is no less mistaken.

The best way to show this is to show how the second type of grasp works. We can do this by focusing on the question of what exactly Galileo's law tells us. It is not merely a description of what happens in one particular circumstance -- fall in a void -- but a description of whatever is common to all cases of free fall. What is especially significant about the case of fall in a void is that the only factors affecting the rate of fall are those which are common to all cases of fall, and fall in a void is unique in this respect since in every other case of fall there are other factors operative which are not common to every instance of fall. Thus we can state the law in two ways, as maintaining that the component of motion common to all instances of fall is a uniformly accelerated motion (call this version *S*1) and as maintaining that all bodies fall at the same uniformly accelerated rate in a void (*S*2). The two statements of the law comprise two different sentences which have different senses. They provide different routes to the same referent. This is an important part of what Galileo establishes. The series of experiments that I described in the last section takes us from *S*1 to *S*2. It takes us from a sentence which has a clear sense, but a sense which provides us with a route to the referent which we are unable to associate with a useable mode of presentation of its referent, and hence we are unable to grasp what we would have to do to determine the truth or falsity of the sentence. What it takes us to is a sentence that we can associate with an empirically determinate mode of presentation. *S*1 is in some ways analogous to the mathematical case that I mentioned above, where we are asked to calculate the value of the function x^2+2x, where x is given in the form of 'the population of Hawaii'. To be able to compute the value of the function we need to be able to translate 'the population of Hawaii' into the canonical form of presentation of, or canonical notation for, numbers, namely numerals in binary notation. Similarly with *S*1. We are helpless if we are merely told that the component of fall common to all instances of fall is a particular kind of motion. If we can show that *S*1 refers to the same situation as *S*2, however, then we can present the law in a fashion which allows its sense to be grasped in such a way that the route to the referent corresponds to a procedure by which the referent can be presented (in this case indirectly) to us. In this way, we provide the law with determinate empirical content.

In sum, *S*1 and *S*2, in virtue of having different senses, have different routes to the referent. We can associate the latter with a set of empirical procedures by which its truth can be determined, since we know that the non-universal features of fall will arise only in fall in resisting media, and the factors operative in this can be determined empirically. In the case of *S*1, however, which is the form of the law naturally suggested by purely conceptual and mathematical considerations, we have no mode of presentation which we can associate with its route to the referent. So what we have to do is to transform *S*1 into *S*2.

The route to the referent indicated by the sense of a sentence is a purely conceptual route, and may completely transcend what we can observe. If all we grasp is the sense of a sentence, if we cannot associate the route to the referent which it provides us with any empirical mode of presentation, then the sentence is not so much devoid of meaning as devoid of empirical content. It is the mode of presentation of the referent that is instrumental in determining whether the sentence has empirical content, or mathematical content, or whatever. The route to the referent supplied by the mode of presentation of the object is, in the empirical case, constrained by such factors as what we can observe, what means of detection we have available, and so on. The aim of the enterprise is twofold: to extend the second type of route as far as we can so that it can join up with the first, and to be able to reformulate the first to such an extent that it can lock onto the second. Experiment has an absolutely crucial role to play in this enterprise, a role which, because it is not simply concerned with confirmation and falsification, has been overlooked. The right theory of meaning enables us to elucidate this role, the wrong ones obscure it.

In elucidating this role, moleculism also allows us to elucidate two connected questions, which may be worth mentioning as they help place what I have argued here in a broader context. The first is the question of realism and instrumentalism. If the argument that I have presented is correct, then certain forms of realism and instrumentalism will be ruled out from the start, namely those forms of realism which work on the assumption that a grasp of sense alone is sufficient for us to attribute determinate empirical meaning to a sentence, and those forms of instrumentalism that consider that a grasp of the means by which a sentence is to be verified is sufficient for a grasp of its meaning. There will, nevertheless, still be substantive disagreements between realists and instrumentalists. They will occur in those cases where mode of presentation and sense provide routes which do not meet, or where the only appropriate modes of presentation go beyond anything we can observe. Such cases are much more common than the prototypical case, such as Galileo's kinematic account of free fall, where complete success is possible. This does not, of course, preclude the prototypical case being prototypical.

The second topic is that of capturing the systematic structure of bodies of theory. Intralinguistic theories of meaning operate with a conception of meaning whereby meanings form a systematic network. This network provides the unifying structure for the body of theory, which then faces the 'tribunal of experience' as a whole. It is at this juncture that the Quine-Duhem thesis, according to which apparent conflicts with evidence can always be accommodated by changes in the network of meaning, comes into play. Every threat to a particular theory is a threat to the whole body of theory on an intralinguistic account because of the theory's systematic meaning relation with the other theories that make up the whole body of theory, thereby creating the need for wholesale revision every time some component of the whole is challenged or threatened. The Quine-Duhem thesis delivers us from this consequence, by allowing us to make adjustments to the network of meaning in response, so that the danger of constant wholesale revision can be averted. But this is achieved at the cost of making every significant connection between theories ultimately one of meaning. This is simply not plausible. As regards the standing of evidence which apparently contradicts some theory or one of its predictions, for example, we surely want to be able to say, at least on occasion (and perhaps often), that a theory is just mistaken. But, as Dummett (1975: 119) has put it so succinctly, holism 'can give no determinate content to the notion of a mistake'. Moreover, there are a number of problems about the inferential connections between parts of a general body of theory which are independent of the meaning relations between those parts. For example, in the sixteenth and seventeenth centuries disputes arose over whether or not physical problems could be posed and resolved mathematically. These centred on the question of whether inferences should take the form of mathematical or syllogistic demonstrations. These were deep disputes that bore on the very nature of physical theory (*cf.* Gaukroger, 1978; 1980), but they have nothing to do with meaning.

In sum, there is a tendency in intralinguistic theories to collapse evidential and inferential relations into meaning relations. No such tendency is present in moleculism, and we are thereby not prevented from capturing the complexity of the structural unity of bodies of theory in a non-reductive fashion. This structural unity is not wholly independent of meaning, but it does not derive exclusively from some single all-embracing network of meaning. It derives equally from shared conceptions of what counts as evidence, shared commitment to ways of formulating problems and resolving them, shared background theories, and so on. This renders the boundaries of bodies of theory much more amorphous and fluid than, for example, on contextualist criteria but then contextualists have occasionally realized this: Kuhn's numerous and often inconsistent ways of characterizing paradigms (*cf.* Kuhn, 1970; Masterman, 1970) fits ill with his contextualism, for instance, but shows sensitivity to the problem of describing how theories are grouped together and unified in historically realistic terms.

4. CONCLUSION

My aim in this paper has been to show that questions of meaning are relevant to understanding the role of experiment, and that such understanding leads us in the direction of moleculism. In particular, I have challenged the view, to which intralinguistic accounts are committed, that there is no role for experiment other than that of determining the truth or falsity of theories developed in some purely conceptual fashion. This is a widely held view, both among advocates of an intralinguistic account of meaning and among those who mistakenly see no role for a theory of meaning in the philosophy of science at all. I have tried to show that it is a false view, and more importantly what its falsity consists in, namely, a failure to recognise that a grasp of meaning involves two different kinds of procedure: one providing us with a conceptual route to the referent, the other providing us with a way in which the referent can be presented to us empirically (in the case of concrete objects). In the case where, for some particular sentence or theory, these fail to enable us to identify the referent we must try to shift to a formulation which picks out the same referent via a different route, one which can be associated with a successful empirical mode of presentation. Experiment plays an absolutely crucial role in this shift, a role which has gone unrecognized because questions of experiment have not been posed in a way which would enable it to come to light.

NOTES

1. I am concerned in this paper only with truth-conditional theories of meaning, and my advocacy of moleculism owes a good deal to Dummett (1981a).
2. I have described this case much more fully (1978: ch. 6), where detailed criticism of Koyre's interpretation of Galileo can also be found. The account that I am offering here differs considerably from my earlier one, however, in that the earlier account hardly touches on questions of meaning and, where it does, it assumes a vague kind of contextualism, which I reject here.

REFERENCES

Davidson, D. (1984). *Inquiries into truth and meaning*. Oxford: Clarendon Press.
Dummett, M. (1975). What is a theory of meaning? In *Mind and language*, ed. S. Guttenplan, pp. 97-138. Oxford: Clarendon Press.
Dummett, M. (1981a). *Frege: Philosophy of language*. 2nd ed., London: Duckworth.
Dummett, M. (1981b). *The interpretation of Frege's philosophy*. London: Duckworth.
Galileo Galilei. (1974). *Two new sciences*. Trans. S. Drake, Madison: University of Wisconsin Press.
Gaukroger, S. (1978). *Explanatory structures*. Sussex: Harvester Press.
Gaukroger, S. (1980). Descartes' project for a mathematical physics. In *Descartes*, ed. S. Gaukroger, pp. 97-140. Sussex: Harvester Press.
Hacking, I. (1982). Language, truth, and relativity. In *Rationality and relativism*, ed. M. Hollis & S. Lukes, pp. 48-67. Oxford: Blackwell.
Hacking, I. (1983). *Representing and intervening*. Cambridge: Cambridge University Press.
Hacking, I. (1985). Styles of scientific reasoning. In *Post-Analytic philosophy*, ed. J. Rajchman & C. West, pp. 145-65. New York: Columbia University Press.
Koyre, A. (1968). *Metaphysics and measurement*. London: Chapman and Hall.
Koyre, A. (1978). *Galileo studies*. Sussex: Harvester Press.
Kuhn, T.S. (1970). *The structure of scientific revolutions*. 2nd ed., Chicago: University of Chicago Press.
Masterman, M. (1970). The nature of a paradigm. In *Criticism and the growth of knowledge*, ed. I. Lakatos & A. Musgrave, pp. 59-89. Cambridge: Cambridge University Press.

University of Sydney

ANDY PICKERING

OPENNESS AND CLOSURE: ON THE GOALS OF SCIENTIFIC PRACTICE

Two decisions: either I recognise the object by its shadow, which gives rise to geometry or, better yet, to the idealism of representation, or I allow for a kernel of shadow within the object. In the latter case, theory and practice develop this secret infinitely in a perpetually open history, the history of science....
The global wandering, the mythical adventure, is, in the end, only the general joining of these spaces, as if the object or target of discourse were only to connect..

—— Michel Serres

Openness and closure are enduring themes in social studies of science. At issue is the insight that knowledge production, and scientific practice in general, is an open-ended and uncertain affair: existing culture does not determine the contours of its own extension. This raises the question of how and why actors, in fact, extend culture in one way rather than another, of how openness is managed, of how closure is achieved, in practice. Many studies bear upon this question, but I want to approach it from a new angle. My aim is to open up for analysis the neglected topic of goal-formation in science.

First I need to take my bearings within the existing literature. The *locus classicus* for discussion of openness is the controversy study (Collins, 1981; 1985; Knorr, Krohn & Whitley, 1981). This genre, invented by Harry Collins in the early 1970s, explores controversies between groups of scientists over matters of empirical fact and theoretical interpretation. Its rationale is that in controversies the contesting parties open up one another's prior closures, pointing to the lack of compulsive force behind, say, controversial facts, and thus demonstrating that, in particular instances, the development of scientific culture could have gone in other directions. In controversies, then, scientists do the analysts' work as far as openness is concerned. But while controversy studies suggest that all closures can be unpicked and reopened given sufficient will and effort, they shed little light on how closures are themselves achieved; on why, for example, scientists seem

H. E. Le Grand (ed.), Experimental Inquiries, 215–239.

to be confident of their particular closures -- novel facts and interpretations -- at least until controversy gets under way.

To get at such questions it helps to move from the rather static frame of controversy studies to the temporally extended frame of ethnographic and ethnomethodological studies (Knorr-Cetina, 1981; Lynch, 1985a & b; Lynch, Livingston & Garfinkel, 1983; Suchman, 1987). These reveal that while closures can always be unpicked they are not always easy to come by. The intentional and pragmatic nature of scientific practice is clearly displayed in ethnographic studies and a somewhat different sense of openness surfaces here, that of a gap between plans and their implementation in the material world. The plan might be, for example, to build a particular piece of apparatus with the goal of measuring some quantity of interest. What emerges from the ethnography of the laboratory is that the implementation of such plans is continually upset by unforeseen blockages, obstacles, resistances encountered in material practice. Much of scientific practice can be understood as the open-ended, creative and embodied work involved in the practical management of resistances; closure can be seen as the achievement of some goal as a successful outcome of this management process (Pickering, 1989).

Ethnographic studies, then, shed light on the day-to-day repair work of laboratory practice that can, contingently, lead to a kind of non-trivial closure that is prior to controversy. One element remains to be analyzed: what of the plans that laboratory practice implements? How should we get to grips with them? Where do they come from? Remarkably little has been said on this subject in the ethnographic literature, or elsewhere within the science-studies orbit for that matter.[1] Here I return to the topic of this essay. I want to analyze the production of plans in science, and to tie this to a further exploration of the themes of openness and closure. To do so I need some conceptual apparatus.

First, for clarity, let me make it explicit that I want to think of scientific research as a two-phase process. One phase is that of the formulation in conceptual practice of *plans*, meaning the formulation of *ends* to be achieved, and of *means* to arrive at them. The other phase is that discussed in the preceding paragraph, that of the *implementation* of plans in laboratory practice. I do not want to suggest that scientists simply formulate unrevisable plans and then seek unreflectively to implement them, but, nevertheless, the following case study will show that there is a useful distinction to be made here. So, my interest lies primarily in the formulation of plans in conceptual practice and I begin with a consideration of the formulation of ends, of goals to be achieved.

A goal is an image of a future state of culture, a state in which some piece of work has been done which has not, in fact, yet been done, an image of something which does not exist. Phrased in this way, we see the problem of openness emerging in a new form, in conceptual practice alone: it seems that anything might be a candidate for a description of such a nonexistent future state. Somehow, though, this rather absolute openness is successfully managed by scientists and I want to talk now about how this management is progressively accomplished. At a very general level, it is clear that scientists do not construct futures from thin air. The future is made from the present and the past in a process of *modelling* or analogy: goals as future states are built out of existing cultural elements, analogically transformed. Thus, membership in, say, a technical culture lends some degree of closure and order to the process of goal formation. But here it is important to remember that analogy itself remains an open-ended concept. Just because everything is similar to, and different from, everything else in some respect, there is no intrinsically right, natural or proper direction for analogy to take. Even for the culturally embedded actor, the problem of the management of openness in goal formation remains.[2]

To proceed, we need to think more carefully about the nature of scientific culture. While it has long been traditional to think of science as having a unitary, monolithic, homogeneous culture; *e.g.*, a single theory or conceptual network, empirical studies suggest that such a view is inadequate. Better, it seems, to see scientific culture as endemically *patchy*, composed of all sorts of *heterogeneous elements* standing, at any given time, in relationships to one another ranging from coherence, through irrelevance to outright contradiction. Scientific communities typically sustain a whole spectrum of conceptual networks, several at any given level of abstraction, a patchy material culture, too, of artefacts and experimental techniques, a range of work-styles, and so on. Within such heterogeneous cultures, the management of openness takes on new dimensions. A whole field of practice, one that could not exist within a unitary culture, comes into view, centred on the possibility of making *connections* between disparate cultural elements and, in the case of experiment, between the material world and elements of articulated culture. This is where new strategies for closure become evident, in a kind of triangulation process. In essence, I suggest that goals situate themselves at imaginatively constructed *intersections* between cultural elements. Such intersections serve further to close down the openness of goal formation: analogy can, in principle, point anywhere, but the establishment of a connection between specified cultural elements requires that it points in one direction rather than another. This idea -- that in goal formation the openness of analogical projection is closed down in reference to the intended establishment of connections within a heterogeneous culture -- is at the heart of this essay.[3]

A few more general remarks are necessary before I proceed to a case study that exemplifies the argument. I spoke earlier of plans as embracing both means and ends, whereas the last few paragraphs have been concerned exclusively with ends. No particular problem arises in reintroducing means into the discussion: as in the case of goals, one can think of means as modelled upon existing cultural elements, one can think of them as establishing connections, now through time rather than across culture, and one can think of the openness inherent in the formulation of plans as being further closed down, as above, in the imaginatively contrived intersection of means with ends. This leads into another idea that will be exemplified in the case study, that of increasing degrees of closure in the piling up of intersections. There is no need to suppose that either means or ends are defined by single models or by a single intended cultural connection. Many intersections can enter into the construction of a plan, each acting further to specify it and to make the future more precise. In what follows, for example, I draw attention to intersections between material objects and systems of representation, the latter serving as the means of temporal transport and transformation of the former, and to the increasing degree of closure implicit in such intersections. I have now repeatedly spoken of degrees of closure, a form of words somewhat at odds with the dichotomous pairing, open/closed. In what follows, I often find it more appropriate to speak of *stabilization*, with its connotation of a continuous range of possibilities, from wildly unstable and ill-defined glimpses of remote futures to finished products (Rudwick, 1988 adopts the same strategy).

I move to a case study intended to exemplify and elaborate on the above analysis: Luis Alvarez's construction of the first large hydrogen bubble chamber. Bubble chambers were perhaps the most important tools of experimental high-energy physics (HEP) from the late 1950s to the mid-1970s. Alvarez's chamber in particular marked an important point of entry of the big-science form of life into HEP experiment.[4]

1. THE GOAL OF A NEW MACHINE

Luis Walter Alvarez was born in San Francisco in 1911. He trained as a physicist at the University of Chicago, where he was awarded his doctorate in 1936. In the same year he joined the 'Rad Lab' -- Ernest Orlando Lawrence's Radiation Laboratory at the University of California, Berkeley -- as a Research Fellow and remained there until his death on 1 September 1988. The move to Berkeley had a formative influence upon Alvarez's research. Although I will not go into details, it is clear, especially from his autobiography (1987), that Alvarez had enormous admiration for Lawrence, both in his person and in his work-style, and that he modelled himself upon Lawrence. For Alvarez, as for Lawrence before him, that meant attempting to outclass the opposition by engaging in large-scale projects on the boundary of physics and engineering: during World War II at MIT and Los Alamos (the opposition being first the Germans, finally the Japanese); after the war in building a large proton linear accelerator at the Rad Lab (the opposition now being Alvarez's fellow particle physicists); and, in the early 1950s, in another military project, the construction of the Materials Testing Accelerator at what became the Livermore National Laboratory (now the enemy was the Soviets). When this last project collapsed in 1952, Alvarez decided to re-enter the world of HEP but found himself in the unusual situation of being out of touch with the research front, of being without a continuing research project, without a goal. That makes Alvarez an interesting site for the exploration of goal formation in science.

At the April 1953 Washington meeting of the American Physical Society (APS), Alvarez met Donald A. Glaser, a physicist from the University of Michigan. Glaser discussed his work on a new kind of elementary-particle detector, the bubble chamber. The bubble chamber, as Glaser described it, was a small container of superheated liquid held under pressure. When the pressure was released the liquid boiled. He had succeeded in demonstrating that boiling began preferentially along the tracks of any charged particles that had recently traversed the chambers. These tracks could be photographed and subsequently analyzed for the information they contained on elementary-particle interactions. In essence, then, Glaser had invented a new kind of particle detector, modelled upon existing cloud-chamber devices, but different from them in containing a liquid rather than a gas as the active medium. Within the existing culture of HEP, the significance of this difference was that particle interactions would take place much more frequently within a dense medium like a liquid than within a gas. The bubble chamber thus represented a potentially more productive experimental device than its predecessors -- more useful than the cloud chamber, for example, in the exploration of rare physical processes. This advantage of the bubble chamber over the cloud chamber was potential rather than actual when Alvarez first met Glaser. Glaser's devices were small prototypes, roughly an inch in diameter, capable of demonstrating the phenomenon of track formation, but by no means developed to the point at which they could rival existing cloud chambers as practical particle detectors. Nevertheless, within hours of hearing of Glaser's work Alvarez *formulated for himself a new goal*: to build a large bubble chamber filled with liquid hydrogen for operation at the Bevatron -- the particle accelerator then under construction at Berkeley, designed to supply the highest energy particle beams in the world (Alvarez, 1968: 117).[5]

I want now to analyze as carefully as I can the formulation and subsequent stabilization of Alvarez's goal and of the plan in which it was embedded. One can begin by noting that although Alvarez's goal of a big bubble chamber referred to a non-existing future state of his field, his choice of goal was by no means completely arbitrary or open. Instead, this future state referred back to Alvarez's present: it contained a bubble chamber like Glaser's but transformed. Most conspicuously, Alvarez foresaw a chamber much larger than any that presently existed. His chamber was, then, *modelled* upon Glaser's. And indeed, it is clear that such modelling should not be thought of as one possible strategy amongst many for reducing openness. Modelling was *constitutive* of Alvarez's new goal, in the sense that it is impossible to envisage his deciding to build a bubble chamber without the example of Glaser's prototype before him. One cannot understand this passage in the history of science, then, without getting to grips with this process of modelling.

So, what more can be said about modelling? First, as already indicated, while modelling has to be seen as constitutive of the stabilization of goals it is nevertheless, itself an open-ended process. Given solely the material example of Glaser's chamber, anything (or nothing) can be said to follow from it. It is possible to imagine any number of variants -- in material construction, shape, size, filling, colour, *etc.* -- and none of these variants has a particular claim to be the true, right, natural or proper elaboration of the object in question. Nevertheless, it would be a mistake to think of Alvarez's projection of the bubble chamber, even in its earliest stages, as being completely free and open. Let me explain why.

At the Washington APS meeting, Glaser did not hold up his actual chamber and invite his colleagues to make of it what they would. Instead, he offered a *description*, a representation of the chamber in words, formulae and diagrams, drained of its materiality and formulated within a range of articulated discourses concerning the physical properties of liquids, the mechanisms of bubble formation, the properties of construction materials, the interactions of elementary particles, and so on. This move from material object to representation is significant in the present context for several reasons. First, it serves to exemplify the theme of *heterogeneity*. The discourses onto which Glaser dematerialized his chamber differed from, and existed largely independently of, one another within his culture. Theories of elementary-particle interactions, for instance, had little to do with engineering characterizations of the materials with which Glaser worked. Likewise, Glaser's material practice itself could not be reduced to any articulated discourse. Nevertheless, in his practice prior to the Washington APS meeting, Glaser had established and stabilized *connections* between all of these cultural elements: his description to Alvarez of his chamber was the fruit of *intersections* contrived in the laboratory between discourse and material practice, of the closure of the gap between his plans for the chamber and their implementation.[6] Furthermore, Glaser's description of his chamber was more than a static representation of what had been accomplished; it was also a means of projecting that accomplishment into the future.[7] This returns us to the topic of modelling.

I said earlier that Alvarez's future chamber was modelled upon Glaser's and it is now possible to be more precise about the nature of modelling. While modelling and analogy are conventionally conceived as primitive operations on similarity and difference, Alvarez's transformation of Glaser's chamber went via a circuit of representation. Glaser described his chamber within, for example, the language of the physical properties of construction materials, marking out certain features of his chamber, like linear dimensions, as significant within that framework, while implicitly backgrounding other attributes, like colour, that fell outside the framework. And thus, in contriving intersections with established discourses, Glaser's description served to close down the openness of the future, to encourage and stabilize some directions for the elaboration of his chamber while leaving others unthinkable or, at any rate, unthought. Hence, under Glaser's description, transformation of the dimensions of the bubble chamber, for example, was an immediately present move, while repainting the laboratory was not. And, as we have seen, Alvarez immediately made the former move rather than the latter, imagining a drastic scaling-up in size.

Here we see, then, how intersections between cultural elements, between Glaser's material practice and various established discourses, served in the practical management of openness, in the stabilization of future goals. But it is important to recognise that this stabilization of Alvarez's vision of the future was only partial, for at least three reasons. First, the discourses Glaser drew upon in the description of his chamber should themselves be seen as conventional and revisable, not fixed forever. Second, Glaser's particular connections of those discourses to his chamber lacked even the aspect of established convention -- indeed, as we shall see in the discussion of the 'dirty' bubble chamber to follow, although Alvarez initially accepted Glaser's description of the chamber, he soon changed it in significant ways. Third because, even if one took Glaser's particular projection, to what extent should one transform the dimensions of the chamber, for example? Why transform dimensions rather than geometry? I will return to the question of dimensions in a moment but first I want to discuss another imaginative transformation wrought on the chamber by Alvarez. For various reasons, including practical convenience and high density, Glaser worked with chambers filled with heavy liquids -- organic chemicals. Alvarez, in contrast, imagined a future chamber filled with liquid hydrogen. Why?

Here it becomes necessary to refer to another line of development in particle detection already under way at Berkeley and elsewhere: the use of small liquid-hydrogen targets in HEP experiment. Such targets were 'passive': they were placed in a particle beam and particle interactions took place within them; these interactions were then monitored by separate detectors placed around the chamber to catch the emergent particles. Again, Alvarez imaginatively transformed the hydrogen target and projected it into the future. He envisaged a bubble-chamber filled with liquid hydrogen as an active target/detector. The chamber would be placed in a beam, and interactions would both take place within it and be made visible there as tracks for recording and analysis. Thus, in formulating his new goal Alvarez projected not one model but two: Glaser's chamber and the hydrogen target. And, moreover, his projected future state lay at the *intersection*, at the point of fusion or *condensation*, of these two projections, further stabilizing his goal: the aim of realizing an active hydrogen target dictated that Glaser's chamber should be projected along one axis -- the transformation of material filling -- rather than others. Note, again, that such intersection or fusion of projected models is only possible within a patchy culture: Glaser's bubble chamber and the hydrogen target were disconnected elements of the culture of HEP until Alvarez brought them together in the context of his project.[8]

I have so far discussed the initial delineation of Alvarez's novel technological goal -- the big hydrogen bubble chamber. We can now turn to the how and the why of the new machine, taking the why first. Why did Alvarez envisage a chamber filled with hydrogen instead of the heavy liquid used by Glaser and why scale it up in size? To answer these questions we need once more to think about representation and about imaginatively contrived intersections within heterogeneous cultures. To reiterate a point made earlier, at the APS meeting Glaser did not simply present the material example of his bubble chamber. Neither did he offer it as a piece of abstract art. Instead, he described his chamber as an elementary-particle detector: a device for making connections between heterogeneous cultural elements, between particle beams in their materiality and the various discourses on the nature of elementary particles sustained within HEP. Alvarez again imaginatively projected the chamber under Glaser's description of it, aiming to capitalize upon and make real this announced potential of the chamber. Alvarez intended to produce data that his colleagues in the HEP community would use, that would be taken up in the practice of HEP theorists, and that would set markers for further experiment. Here, the choice of a hydrogen rather than heavy-liquid filling was significant. Within the conventional discourse of the time, hydrogen was agreed to be the simplest and most readily interpretable target for HEP experiment. Data taken on hydrogen, with its nucleus of a single proton, called for none of the complicated approximations used to extract information on the properties of elementary particles from data taken on complex nuclei, with the inherent loss of precision that such approximations entailed. Hydrogen, from this perspective, was the best choice of target for HEP experiment. Since, all things being equal, the data-production rate of an HEP experiment is proportional to the mass of its target, Alvarez's big hydrogen chamber represented an attempt to produce and circulate large quantities of high-grade information. Alvarez's projection, in other words, situated and stabilized itself at yet another intersection, now between its product -- photographs of elementary-particle interactions -- and the various contemporary discourses on the nature and properties of such interactions.

There were, however, countervailing reasons not to use hydrogen. Hydrogen boils, for instance, at a temperature close to absolute zero. Use of liquid-hydrogen targets therefore required the importation of sophisticated cryogenic techniques that were unnecessary for handling Glaser's heavy liquids -- another instance of establishing intersections between the heterogeneous elements of the culture of physics. The already existing hydrogen targets demonstrated to Alvarez that such technical problems could be overcome, at least for small targets.

Another reason not to use liquid hydrogen as a target was its low density. As mentioned above, low density implied that a hydrogen chamber would have to be proportionately larger than a heavy-liquid chamber to achieve the same productivity. This was, in part, the reasoning that led Alvarez to conceive of a chamber much larger than Glaser's. There was, however, another line of reasoning that led to the same conclusion. Scaling up the chamber, together with the cryogenic techniques just mentioned and the safety problems attendant upon working with large quantities of a highly explosive medium like hydrogen, meant that the construction of Alvarez's projected chamber would be an expensive and difficult engineering project. It appears that most of the experimenters who knew of Glaser's work at this stage saw this scenario as problematic, more of a *destabilizing clash* between projections of the bubble chamber and the hydrogen target than a stabilizing intersection. Alvarez took the latter view. To understand why, one needs to refer again to intersections within a heterogeneous culture and back to Alvarez's biography.

HEP in the early 1950s sustained two different work-styles. One was the small-science style -- individualistic, requiring small amounts of funding, the typical bench-top approach that had characterized experimental physics throughout its history. This style continued to dominate experimental HEP and it was within this frame that the engineering connotations of building a big hydrogen chamber constituted a destabilizing clash. The other work-style was, of course, the big-science approach -- characterized by joint efforts of physicists and engineers, hierarchically organized into teams, consuming large amounts of funding, and so on. This was the style that had been developed by Alvarez's mentor, Lawrence, at Berkeley in the 1930s, that had proved so effective in the US weapons laboratories in World War II and after, and that had come to dominate the postwar HEP accelerator-building culture. It was also the style in which Alvarez had been immersed for more than a decade when he met Glaser. Large-scale projects on the boundary between physics and engineering thus hung together with aspects of Alvarez's background and expertise as they did with few other physicists'. As became clear once the project got under way, in constructing his chamber Alvarez was in a position to *model* his practice, in dragging the present state of HEP into his projected future, upon what he had done successfully several times before in capital-intensive, hierarchically-organized, big-science projects at Berkeley, MIT, Los Alamos and Livermore. Alvarez's projected chamber thus situated itself at yet another intersection of heterogeneous elements, this time of the big-science work-style with Glaser's bubble chamber and of the hydrogen target, both imaginatively transformed and condensed. This additional intersection served further to reduce the openness of Alvarez's goal. The dimensions of the chamber now had to lie within the realms of the engineering expertise, organization and resources that Alvarez could mobilize.[9]

Besides this reduction of openness, the intersection of the big-science work-style with the future chamber had a further significance: it was an intersection of *ends and means*. It meant that Alvarez had a vision both of a future state of HEP and of how to get there. It provided Alvarez's future chamber with a kind of imaginary concreteness, with an imagined history leading up to it from the present. With the addition of the big-science element, then, Alvarez's goal became integrated into what I earlier called a *plan*, and all that remained for him was to attempt to translate his plan into reality, to *find out* if he could build a hydrogen bubble chamber at all, and, if so, to what limits of gigantism he could take its development.[10] This proved to be a major undertaking extending over many years. In the context of the present discussion I need to talk about some aspects of it but not others. Thus, I will not attempt to enter into the day-to-day dialectic of resistances and accommodations entailed in closing the gap between Alvarez's plan and its implementation, since this kind of process is already an active topic for ethnographic enquiry. Suffice it to say that this dialectic played a continuous role in delineating the final form of the chamber up until the moment of its completion; that Alvarez's goal was never, in this sense, precise and fully stable until it had been achieved.[11] Following on from this, however, I do intend to emphasise two points in respect of the history of the construction of Alvarez's big chamber. First, despite the piling-up of intersections that determined the relative closure and stabilization of his initial plan, that plan itself was still vague and open-ended to some degree. It remained to specify it more closely in practice, in interesting ways that I shall discuss. And second, I want to stress that even aspects of Alvarez's initial plan that were relatively precise were not unrevisable.

2. BUILDING THE BUBBLE CHAMBER

One can think of the plan with which Alvarez returned from the Washington meeting to Berkeley as a set of integrated and mutually stabilizing maps of the present future state of the culture of HEP, maps of what his big hydrogen chamber would look like, of how it would be used and of what resources needed to be mobilized to build it. All of these maps should be thought of as corrigible and revisable, as open to further specification. Some of them were ready to be tried out in practice and proved to be adequate to Alvarez's purposes, others required further work of one kind or another. It is the latter set of maps that interests me most here but I comment briefly on the former.

Part of the attraction for Alvarez of the large bubble chamber was its big-science aspect: he did not intend to build the chamber alone at the laboratory bench. Integral to his plan was the mobilization of the unparalleled technical and engineering resources of the Rad Lab -- of expertise and material -- in pursuit of his goal. In this respect he seems to have encountered little resistance. Several physicists and technicians at Berkeley quickly set to work on bubble chamber construction, first successfully following Glaser's steps in building small heavy-liquid chambers and then moving on to the novel task of attempting to build small hydrogen chambers.[12] We return to this early construction work shortly but first note that this success in garnering resources marked Alvarez's entire project. At no point, either during the construction of the large chamber or in its subsequent period of use, does Alvarez seem to have experienced significant difficulties in drawing increasing numbers of the Rad Lab staff into the project. Likewise, when the project had reached the point of requiring major outside funding, few obstacles stood in Alvarez's way. After a day of lobbying in Washington by Alvarez and Lawrence, the US Atomic Energy Commission (AEC) agreed to fund the project at the level of $750,000, and later never balked when the final budget ran to $2,500,000. Alvarez's maps of resources to be mobilized proved to be reliable -- and given his extensive practical engagement with both the Rad Lab and the AEC this is hardly surprising.

I now turn to some instances where Alvarez's maps did require further work of some kind, starting with the history of bubble chamber construction at Berkeley. From May 1953 onwards, physicists and technicians at Berkeley struggled with all of the problems in constructing chambers that Glaser had encountered before them, working first with heavy liquids and then moving on to liquid hydrogen, finally producing a prototype hydrogen chamber that successfully recorded tracks in January 1954. The path was clear to the big machine. Or was it?

One technical problem presented itself in scaling the bubble chamber to larger dimensions. I talked earlier about the descriptions under which Glaser had dematerialized his chamber and through which

Alvarez had projected it. One element of that description needs to be singled out here: the idea that the interior of a successful bubble chamber had to be extremely clean. Glaser's experience was that track formation was only possible if the interior walls of the chamber were of finely polished and respectfully treated glass. Otherwise, upon expansion of the chamber, boiling would begin immediately at the walls, swamping any tracks that might be produced in the body of the liquid. The Rad Lab group had accepted this item of bubble-chamber lore and had verified it in their own experience, laying great emphasis upon cleanliness in their early trials. This emphasis on cleanliness cast doubt upon Alvarez's project. Using standard engineering knowledge to scale up Alvarez's dematerialized prototype, it seemed unlikely that a clean hydrogen chamber orders of magnitude larger than Glaser's could perform as required: glass walls could not satisfy the thermal and mechanical performance requirements.

Here, then, was a *destabilizing clash* rather than a stabilizing intersection of projections, a *situated resistance* encountered in conceptual practice.[13] The future chamber that Alvarez saw no longer intersected with existing chambers projected under an existing description. Something had to give. It might have been the plan itself: Alvarez might have concluded that the construction of a big hydrogen chamber should be abandoned. Contingently, the description itself gave way. I noted above that Glaser's description of his chamber, though contributing to the relative closure of future extensions, should not be though of as fixed and unrevisable. The Berkeley group found a way to revise it, transforming the clash of projections into a new intersection that further stabilized and defined Alvarez's goal. It is interesting that this circumvention of a potentially destabilizing resistance was accomplished first in material practice rather than in thought. One of the Berkeley technicians, A.J. Schwemin, simply went ahead and built a relatively large, 2.5-inch diameter, dirty hydrogen chamber, a chamber with plane glass windows mounted on a cylindrical metal body. It was found, as expected, that boiling did set in at the metal-glass joints when the chamber was expanded but John Wood, another technician, observed that, even so, tracks could still be discerned in the body of the liquid. Photographs of tracks could thus be taken even in a dirty chamber -- meaning one with metal walls -- providing the optics and timing were adequate to distinguish bubbles in the central volume (Alvarez, 1968: 118-9). Furthermore, since the ratio of volume to surface area grows in proportion to linear dimension, residual problems connected with dirty walls could be expected to diminish in significance as chambers became larger. Returning to engineering knowledge of materials, there appeared to be no problem in constructing a large metal-bodied chamber using glass only for viewing ports. The path to the big hydrogen chamber was once more clear but *under a new, revised description* and Alvarez now had a new, more precise and more stable goal: a large metal-bodied chamber.[14]

Work now proceeded at Berkeley along two fronts. A series of larger, metal-bodied hydrogen chambers were constructed. By December 1954 the phase of prototypes and demonstration models was over: a four-inch chamber was in use for experiment and design of a ten-inch chamber, the penultimate step on the way to the Alvarez's big machine, was under way. At the same time, Alvarez was refining his vision of what that machine should be. This returns us to consideration of the vagueness, the residual openness, that remained in his plan even after he had set out to implement it. Many questions concerning just what the technical specifications of the big chamber should be went initially unanswered. He worked upon several fronts, all of which involved the more precise designation of connections within a heterogeneous culture, to resolve them.

I have already been alluded to one front on which Alvarez worked, that of establishing a link between his project and an otherwise disconnected institution. The institution was the AEC; the connection was the flow of funds. I mentioned that Alvarez was successful in this respect but it is important to note the scale of his success evidently set an upper limit on what 'bigness' could mean in practice and helped further to specify his goal. Besides direct funding, the AEC also supported Alvarez indirectly via the AEC-National Bureau of Standards laboratory at Boulder, Colorado. The Boulder laboratory had developed considerable expertise in working with large quantities of liquid hydrogen (in early work on a 'wet' hydrogen bomb) and helped Alvarez's group cope with projected technical and safety problems connected with the big chamber. Collaboration with Boulder achieved yet another intersection between disparate discourses and practices, stabilizing further aspects of the technical design of the chamber and the working procedures implicit in its use.[15]

In parallel with the search for funds, Alvarez set out to produce a detailed design for his big chamber. It was time to move from a vague goal toward a precise list of specifications. The topic then of greatest interest in the HEP community was the nature and properties of the so-called 'strange particles'. Experimenters and theorists were organizing great efforts to find out about and make sense of these particles. Alvarez sought to optimise the parameters of his big detector for insertion into this particular discourse. The strange particles of most interest were neutral ones, hard to investigate experimentally since they leave no tracks in any kind of detectors. Alvarez reasoned that he should therefore make his chamber large enough to contain both the production of such neutral particles and their decay into other, charged, and thus visible, particles. Using known estimates of the lifetimes of the strange particles and taking into account the energy of the Bevatron's beams, he concluded that a length of around fifty inches would be sufficient for his purposes. Leaving a margin for the uncertainties of present and future physics, he worked out a detailed design for a chamber 72 inches in length. This design work -- drawing upon knowledge of strange particles and the parameters of the Bevatron -- further specified and stabilized Alvarez's goal at a relatively precise point of intersection between the machine and the discourses of HEP.

One final development along the way to the 72-inch chamber can be mentioned here. With the design in hand, it is was straightforward to estimate the data-production rate of the chamber; here another destabilizing clash threatened to emerge. It was clear that existing methods of measuring and analyzing photographs of elementary-particle interactions were inadequate to handle the output of the future chamber: the machine would produce photographs much faster than could be converted into useful data. Once more, Alvarez succeeded in defusing this clash and transforming it into a stabilizing intersection. Drawing upon his own expertise in optics and radar and upon developments in computing, he set about devising a system for automating much of the analysis of the 72-inch film. Once again, the contrivance of intersections between heterogeneous cultural elements further stabilized, and elaborated, Alvarez's goal.[16]

This is as far as I want to take the story of the construction of Alvarez's big bubble chamber. In the latter half of the 1950s it took on a material form conditioned by the various present and future intersections of cultural elements just discussed. In 1959 it was completed and came into service for HEP experiment at the Berkeley Bevatron, where it enjoyed a long and successful career: data produced by the chamber circulated freely through the HEP community and contributed to many important developments. Alvarez won the 1968 Nobel Prize for physics for achieving the goal that he had set himself at the APS meeting fifteen years earlier. I cannot discuss the uses to which the chamber was put in any detail here. Instead, I close this account with one comment on the discourses into which Alvarez's chamber was inserted.

Alvarez had aimed his chamber at strange-particle physics and it did indeed serve its purpose there. But, the historical significance of the chamber rests less upon this intersection with an established discourse than with the intersection of the chamber with a quite new and unanticipated discourse. The late 1950s saw the beginning of the 'population explosion' of strongly-interacting particles collectively known as 'resonances'.[17] These were particles so short-lived that they decayed to other particles before they had chance to leave any visible track in a bubble chamber. One might therefore assume that the 72-inch chamber, deliberately optimized for the detection of long-lived strange particles, would have had no special advantage in the investigation of resonance physics. But, in fact, the 72-inch intersected centrally with this new field. The experimental investigation of resonance particles quickly came to revolve around the extraction of their properties through statistical analyses of very large numbers of events (beam-particle interactions) and by virtue of its sheer size, large numbers of events were just what Alvarez's chamber generated. Thus, designed to intersect with the discourse of strange-particle physics, the 72-inch chamber acceded to a quite unintended role in the very different discourse of resonance physics.

This switch serves, I think, to make two points. First, as is evident throughout this account, the various maps, projections and intersections implicit in scientific practice have to be understood as actor's constructions, not literal readings of the material and social world. They are therefore corrigible in practice. Second, correction in practice need not take the form of resistance. Nothing stood in the way of Alvarez's projected intersection of his chamber with the discourse on strange particles. Instead, a new and unexpected *opportunity* opened up before him, the emergent field of resonance physics. A notion of opportunism has been lurking in the background throughout this discussion of goals and plans and I return to it shortly.

3. ON PRAGMATISM

I have organized the preceding account of Alvarez and his bubble chamber around the themes of openness, closure and stabilization. I want to close by setting the same history in a broader context. It seems to me that this study contains some key elements of a relatively well elaborated pragmatist model of scientific practice (see also Latour, this volume).

Classical pragmatism, as laid out by William James (1978), insists on the following: that practice is goal-oriented, that goals are pursued within a framework of taken-for-granted procedures and understandings, and that those procedures and understandings are subject to trial-and-error revision in the face of situated resistances. This is the same picture that emerges from the story of Alvarez and the bubble chamber. Alvarez, for instance, understood and projected the bubble chamber under a set of descriptions and understandings inherited from Glaser. Until, that is, Glaser's emphasis on cleanliness proved to stand in the way of Alvarez's goal of a big chamber. The ascribed significance of cleanliness then became subject to pragmatic revision (in material practice, in the first instance) and the set of relevant descriptions was appropriately modified. Other resistances; *e.g.*, difficulties in handling liquid hydrogen and in processing large amounts of data, were imaginatively rehearsed before they were encountered in material practice but were likewise accommodated in revised and elaborated descriptions of the future chamber.[18]

The dialectic of accommodation to resistances evident in Alvarez's practice thus fits well with the pragmatist schema. Classical pragmatism has, however, little to say on the constitution of the goals against which resistances are manifest. Typically, pragmatist thought has concerned itself with routinized goals of daily life. James liked to talk about the man lost in the forest, whose goal is to remain alive by finding food and shelter. Goals like that seem to require little analysis or comment. But in the case of science, where goals are not routine, one needs to do better. That is what I have attempted here. I have sought to explain how new goals and plans are constructed, in terms of an analysis of modelling and of the intersection of elements within a heterogeneous culture. In so doing, while sustaining the classic schema of accommodation to resistances, I have described what I take to be a more general pragmatist framework for the analysis of scientific practice.[19]

One point needs to be emphasized concerning the understanding of goal formation that emerges from this study. The fundamental image is of the scientist imaginatively contriving intersections and connections within a complex cultural space and it is, I think, important to stress that such intersections are made *opportunistically* -- pragmatically, in the everyday sense of the word. It is implicit in my discussion of projection through modelling, analogy and representation that there is no rule, method or algorithm available to scientists for the construction of goals. The same comment applies to the elaboration or revision of goals and plans in the face of resistances, or in the face of remappings of cultural space -- as in the repositioning of Alvarez's bubble chamber with respect to the new field of resonance physics. All such work has to be seen as opportunistic, and this opportunism is the counterpoint to the situated nature of the resistances that arise in the implementation of plans.[20]

Let me close with some remarks on the relation between the general and the particular. As summarized so far, my analysis is intended to elucidate a universal structure of scientific practice. All scientists go on in this way, is the claim: nothing in the framework as yet refers to any particular scientist. But working with the image of a heterogeneous and patchy culture it is easy to begin, at least, to particularize the discussion. Start with the idea that constructing goals requires the projection of multiple intersections between cultural elements, or, to be more precise, between actors' representations of those elements. Since scientific culture is not unitary and uniform, each actor encounters it differently. And therefore the representations of cultural elements, which define the space of goal formulation and practice in general, themselves differ from one actor to another. The possibility for the extension of a pragmatist understanding to the practice of individual actors thus becomes evident. This is not the subject matter of the present essay, but let me carry the analysis a little further, in order to connect it back to the case study.

For simplicity, consider just two ways in which individuals are marked out by their encounters with scientific culture. One is in the painstaking acquisition in practice of relatively immobile cultural elements, like social and technical expertise. Another is in the contingent acquisition in social interactions of more mobile elements, such as knowledge articulated within shared discourses. These different markings are sufficient for a crude explanation of why Alvarez conceived his plan for a giant hydrogen bubble chamber and of why it was Alvarez rather than, say, Glaser who did so.[21] The space in which Alvarez constructed his goal was defined simultaneously by his knowledge of the bubble chamber -- deriving, as it happens, from a chance lunchtime chat with Glaser himself -- and by his hard-won and relatively unique expertise in big science. No other actor moved in the same space; within that space, it is not surprising that Alvarez constructed the goal that he did. To be more precise on the first point, of the small number of people who had heard of Glaser's bubble chamber in early 1953, an even smaller number shared Alvarez's big-science background. On the second point, two remarks are appropriate.

First, I want to emphasize that I am not pointing here to a causal explanation of Alvarez's process of goal construction (at least, as I understand the nature of causal explanation). Undoubtedly, the space of representations in which Alvarez worked was much more complicated than I have described it and I do not want to assert that within that complex, multidimensional space only one goal -- the big hydrogen chamber -- could have been constructed: any analysis that emphasizes the opportunistic aspect of practice would preclude this. I am trying, rather, to sketch an analysis of how a space in which such a goal was constructible came into being, and of why, once that goal had been constructed, Alvarez found it attractive.[22] To clarify what is at stake here, it might be useful to contrast my present analysis with another style of explanation which is in many ways simpler -- explanation by appeal to *interests*. Alvarez, one might want to say, had an evident interest in, for example, doing big science. In one sense, as indicated above, this seems to me to be an unobjectionable formulation: it points to an important attribute of Alvarez as a historical actor. But, in another sense it is deeply misleading. Interest-explanations can easily take on closed form. They lead one to imagine, in this instance, that one element of Alvarez's practice was unrevisably and precisely fixed in advance -- his interest in big science -- while all of the other elements implicit in the bubble-chamber story was subsidiary to it, contrived and stabilized around it. The intrinsic openness of practice disappears on such an understanding, not least the openness of notions like 'doing big science', 'doing experiment' or 'doing strange-particle physics'. These notions, any or all of which might be appealed to as explanatory interests, are just as open-ended as the material example of Glaser's bubble chamber, and just as much in need of practice management and definition.[23]

My second remark is this. Even if I elaborated what I have only sketched out, Alvarez's practice would still not be fully explained. One should think of goals and plans as born in a sea of uncertainties and of stabilization as inherently incomplete. Alvarez had no way of knowing in advance, for example, that he would succeed in recruiting AEC funding for his project, or that he could overcome the problems inherent in scaling up a clean chamber. At any stage he could have failed. Some balancing of investment and return -- some calculus of success and failure, risk and reward -- must come into play here in the evaluation of both goals and continuing projects and needs to be understood before the history of Alvarez and the 72-inch chamber is explained in its specifics. Consideration of such balancing acts is beyond the scope of this essay: I have tried to illuminate the stage on which they are performed.

ACKNOWLEDGEMENTS

Versions, or parts of versions, of this essay have been presented at various meetings and seminars, including the joint meeting of the British Society for the History of Science and the History of Science Society, Manchester, England, 11-15 July 1988; the Center for the Study of Science in Society, Virginia Polytechnic Institute and State University; the Sociology Department, Boston University; the History, Philosophy and Sociology of Science Seminar at the University of Illinois; and in my course on the Sociology of Scientific Knowledge at the University of Illinois, Fall 1988. I am grateful for helpful comments received on all of these occasions.

NOTES

1. Knorr-Cetina (1981) develops Kuhn's (1970) explication of goals in terms of a perceived 'promise of success'. This essay takes the analysis further.
2. On the centrality of modelling and analogy to scientific practice, see Barnes (1982), Hesse (1966), Knorr-Cetina (1981), Kuhn (1970) and Pickering (1981; 1984; 1987). Knorr-Cetina explicitly connects modelling with goal-formation; Barnes' analysis of the openness of analogy is especially clear.
3. A unitary conception of scientific culture pervades many contemporary understandings of science. Beyond positivist philosophy, which takes the single scientific theory as its focus of attention, it surfaces, for example, in Barnes (1982), Collins (1985), Feyerabend (1978), Kuhn (1970) and Quine (1980). For works that point to the importance of recognizing the heterogeneity of scientific culture, see Duhem (1954), Hacking (1983), Knorr-Cetina (1981), Pickering (1986, 1989), Pinch (1985) and the literature on the 'actor-network' approach: for example, Michel Callon (1980), Bruno Latour (1987) and John Law (1987). The actor-network approach lays special emphasis on the centrality to scientific practice of the establishment of connections between heterogeneous cultural elements. Gaukroger (this volume) also points to a kind of triangulation process in conceptual practice, though in a different context from that which concerns me here.
4. In what follows I rely upon the extensive secondary literature of Alvarez and the bubble chamber: Alvarez (1987), Galison (1985), Glaser (1960), Heilbron, Seidel & Wheaton (1981) and Trower (1987, forthcoming). Little of what follows is in dispute: I cite the literature only to point to key sources on particular topics. I thank Michael Riordan for suggesting the following heading.
5. Alvarez was apparently searching for some new detecting device to use at the Bevatron even before his meeting with Glaser (Peter Trower, personal communication).
6. In Pickering (1989) I discuss the gap between material objects or practices and descriptions of them, and I analyze the bridging of this gap in terms of a concept of 'interactive stabilization'.
7. For recent studies of representation, see Lynch & Woolgar (1988). Unfortunately for present purposes, these studies highlight the statics rather than the dynamics of

representation: they explore how representation is accomplished, rather than how it serves as a medium of projection and imaginative transformation.
8. Alvarez was accompanied at the Washington APS meeting by a graduate student, Frank Crawford, who was already carrying out liquid-hydrogen experiments at Berkeley with a second student, Lynn Stevenson (1987: 105). Stevenson and Crawford became two of the first members of Alvarez's bubble-chamber group at Berkeley (Alvarez, 1968: 117).
9. Note that this intersection served to close down not only the future shape of the bubble chamber, but also what the notion of 'doing big science' might entail in experimental HEP. Until Alvarez embarked upon his project, this latter notion was completely problematic: there were no examples of it. There is an important symmetry here: all of the elements brought into intersection become better defined.
10. It may be useful to summarize here the various projections and cultural intersections that I have discussed above. These were: the intersections of material artefacts and discourses involved in the projection of the bubble chamber and the liquid-hydrogen target; the condensation of these projections into the big hydrogen bubble chamber; the imaginative positioning of the chamber as a translation device between the Bevatron particle beams and theoretical discourses on the nature of elementary particles; and the intersection of the big hydrogen bubble chamber as end with the big-science work-style as means.
11. Weiss (forthcoming) is the best source on the immense practical difficulties involved in constructing large hydrogen chambers, though her concern is with the chambers built at the European HEP laboratory, CERN, and not with Berkeley. Her study also makes clear how hard it was for the Europeans, lacking the appropriate experience, to adopt the big-science work-style.
12. Here it is interesting to note that in a patchy, localized culture *not all resources are equally mobile*. While words and texts can sometimes flow freely from place to place and individual to individual, other resources, like skills and craft knowledge, are hard to move around. When in 1953 Alvarez set out on his project, the tacit skills of bubble-chamber construction were situated in just one man, Donald Glaser. Alvarez needed such skills if he were himself to build a chamber. One route to their acquisition would have to acquire Glaser himself, to induce Glaser to come to Berkeley, but Alvarez followed the different route indicated in the text.
13. See Pickering (1989) for discussion and exemplification of the concept of situated resistance. The key idea here is that of an interdependence between obstacles -- resistances -- and goals: the former can only exist in relation to the latter, and are *discovered* in practice.
14. Galison (1985: 326) notes another axis along which the bubble chamber was redescribed. In the light of Glaser's experience in the development of a liquid-xenon chamber, Glaser's original theoretical analysis of bubble formation -- which lay at the centre of early chamber design -- was abandoned in favour of the 'heat-spike theory' of Frederick Seitz.
15. Again one can note the symmetry and reciprocity of closure here. With the collapse of the wet-bomb project, workers at Boulder had been in the same position as Alvarez at the demise of the MTA, and collaboration with Berkeley served to refine and specify the mission of both laboratories at once.
16. The importance of Alvarez's background in optics and radar to this initiative in data-processing has been emphasized to me by Peter Trower (personal communication), though more research would be needed to make the connection precise. The emergence

of automated data-processing as a *subgoal* of the bubble-chamber project may be typical of the evolution of a great deal of scientific practice.

17. More and more such particles were discovered through the 1960s, and HEP theorists and experimenters devoted increasing efforts to exploring and understanding their properties -- efforts that led in 1961 to the 'Eightfold Way' symmetry classification of elementary particles and in 1964 to the 'quark' concept that has since become central to particle physicists' understanding of their subject matter. Resonances were thus at the heart of several major discourses within HEP in the 1960s. See Pickering (1984) for details.

18. The open-ended redescribability of material and conceptual objects deserves more explicit attention than I can give it here. See Pickering (1989) for another study of scientific practice that bears upon this issue and, for suggestive work in the history of mathematics, Lakatos' (1976) study of the development of Euler's theorem and Stephanides' (1988) study of the invention of quaternions.

19. In its emphasis on goals, the present essay is intended to complement the analysis of resistance and accommodation of Pickering (1989).

20. The opportunistic quality of the dynamics of scientific practice is stressed in Knorr-Cetina (1981) and Pickering (1984). The latter work underplays the themes of resistance and accommodation.

21. Once Glaser's demonstration of the bubble-chamber principle became known, many physicists around the world set about constructing a wide variety of chambers, though none were as ambitious as Alvarez (see Galison, 1985 and, for Europe, Weiss, forthcoming). I can see no reason why the kind of analysis of Alvarez's practice sketched out here should not be extended to the work of these other physicists.

22. The following suggestion might help to illuminate this remark. Suppose that scientists are continually imaginatively transforming and juxtaposing cultural elements, rehearsing future states and ways of achieving them. The typical output of this process is inconsequential, but occasionally what distils out and becomes a focus for systematic reflection is a plan -- some kind of minimal set of maps of where to go and how to get there. Once this set of intersecting maps is in place it can be refined, augmented and, perhaps, put into practice. This is a very crude model of the process of scientific creativity -- it would be nice to do better -- but it is a sufficient basis to make it possible to address the problem of explaining what is special about the practice of particular scientists. Usher's (1954: ch. 4) gestalt model of technological innovation has much in common with that outlined here, though it lacks the crucial dynamic element of modelling (see also Koffka, 1935). Amann & Knorr-Cetina (forthcoming) can be read as an empirically based elaboration of this kind of model.

23. On the openness of notions like 'doing big science' see notes 9 and 15 above. One of the most plausible articulations of historical causality is that a 'cause' is any element in the absence of which a historical event would not have taken place. But the open-ended modification of the elements that enter into cultural intersections makes even this notion of causality hard to sustain: the contours of a 'cause' may well only be (relatively) clear *after the fact*, even though the cause retains the same name. For classic statements and empirical exemplification of the interest model, see Barnes (1977; 1982: ch. 5) and Shapin (1982). See also the critical exchange between Barnes (1981) and Woolgar (1981a & b), and Latour's (1987) somewhat ambiguous critique of 'interest' (it seems to me that Latour's critique draws attention to the overly closed character of interest explanation, as discussed above). Bloor (1978: 266) and Shapin (1988: note 12) seek to defend the usefulness of 'interest' as an explanatory category but both concede that it stands in need of further explication. Bloor argues that explanatory terms should find their meaning

within some overarching theoretical framework. Shapin, interestingly, articulates his notion of interest in terms of a prior conception of expertise that promises to be consonant with the analysis offered here. But the tenor of his remarks, especially on actors' defence of investments in particular forms of expertise (1988: 545), fails to do justice to the openness of practice that I seek to foreground. Two further points can be discussed in this connection. First, it might be argued that 'interest' is properly a macrosocial concept, not to be elucidated in microsocial analyses like mine. It is hard to see how this argument might go through if one recognises any relation between interest and expertise but, independently of that, my feeling is rather that the macro needs to be reconceptualized in the light of detailed microanalyses -- a feeling well articulated, for example, in Latour (1987) -- and that much still remains to be done in this direction. One needs to think especially about the endurance and circulation of elements within a patchy culture. Latour (1987) addresses such questions, and I offer some preliminary thoughts in Pickering (1988; forthcoming). The second point concerns whether one should aim at a distinctly *sociological* explanation of practice, as adumbrated, for example, in the interest model. Here again I agree with Latour (1987): recognition of the heterogeneity of the space within which actors move precludes any purely social -- or material, or rational, or whatever -- explication of scientific practice.

REFERENCES

Alvarez, L.W. (1968). Recent developments in particle physics. Nobel lecture, reprinted in Trower (1987), pp. 110-53.

Alvarez, L.W. (1987). *Alvarez: Adventures of a physicist.* New York: Basic Books.

Amann, K. & Knorr-Cetina, K. (forthcoming). Thinking through talk: An ethnographic study of a molecular biology laboratory. To appear in *Knowledge and society: Studies in the sociology of science, past and present*, Vol. 8, ed. L. Hargens, R.A. Jones & A. Pickering. Greenwich, CT: JAI Press.

Barnes, B. (1977). *Interests and the growth of knowledge.* London: Routledge & Kegan Paul.

Barnes, B. (1981). On the 'hows' and 'whys' of cultural change (Response to Woolgar). *Social Studies of Science*, **11**: 481-98.

Barnes, B. (1982). *T.S. Kuhn and social science.* London: Macmillan.

Bloor, D. (1978). Polyhedra and the abominations of Leviticus. *British journal for the history of science*, **11**: 245-72.

Callon, M. (1980). The state and technological innovation: A case study of the electrical vehicle in France. *Research policy*, **9**: 358-76.

Collins, H.M., ed. (1981). *Knowledge and controversy: Studies of modern natural science.* Special issue of *Social studies of science*, **11** (1).

Collins, H.M. (1985). *Changing order: Replication and induction in scientific practice.* Beverly Hills: Sage Publications.

Duhem, P. (1954). *The aim and structure of physical theory.* Princeton: Princeton University Press.

Feyerabend, P.K. (1978). *Science in a free society.* London: New Left Books.

Galison, P. (1985). Bubble chambers and the experimental workplace. In *Observation, experiment, and hypothesis in modern physical science*, ed. P. Achinstein & O. Hannaway, pp. 309-73. Cambridge: MIT Press.

Glaser, D.A. (1960). Elementary particles and bubble chambers. Nobel lecture, rept. in *Nobel lectures: Physics, 1942-1962*, pp. 529-51. Amsterdam: Elsevier 1964.

Hacking, I. (1983). *Representing and intervening*. Cambridge: Cambridge University Press.

Heilbron, J.L., Seidel, R.W. & Wheaton, B.R. (1981). *Lawrence and his laboratory: Nuclear science at Berkeley, 1931-1961*. Berkeley: University of California Press.

Hesse, M.B. (1966). *Models and analogies in science*. Notre Dame: University of Notre Dame Press.

James, W. (1978). *Pragmatism and the meaning of truth*. Cambridge: Harvard University Press.

Knorr-Cetina, K.D. (1981). *The manufacture of knowledge: An essay on the constructivist and contextual nature of science*. New York: Pergamon.

Knorr, K.D., Krohn R. & Whitley, R.D., eds. (1981). *The social process of scientific investigation. Sociology of the sciences, Volume IV, 1980*. Dordrecht: Reidel.

Koffka, K. (1935). *Principles of gestalt psychology*. New York: Harcourt, Brace.

Kuhn, T.S. (1970). *The structure of scientific revolutions*. 2nd ed., Chicago: University of Chicago Press.

Latour, B. (1987). *Science in action*. Cambridge: Harvard University Press.

Lakatos, I. (1976). *Proofs and refutations: The logic of mathematical discovery*. Cambridge: Cambridge University Press.

Law, J. (1987). On the social explanation of technical change: The case of the Portuguese maritime expansion. *Technology and culture*, **28**: 227-52.

Lynch, M. (1985a). Discipline and the material form of images: An analysis of scientific visibility. *Social studies of science*, **15**: 37-66.

Lynch, M. (1985b). *Art and artifact in laboratory science: A study of shop work and shop talk in a research laboratory*. London: Routledge and Kegan Paul.

Lynch, M., Livingston, E. & Garfinkel, H. (1983). Temporal order in laboratory work. In *Science observed: Perspectives on the social study of science*, ed. K. Knorr-Cetina & M. Mulkay, pp. 205-38. Beverly Hills: Sage.

Lynch, M. & Woolgar, S., eds. (1988). Special issue on 'Representation', *Human studies*, **11**.

Pickering, A. (1981). The role of interests in high-energy physics: The choice between charm and colour. In Knorr, Krohn & Whitley (1981), pp. 107-38.

Pickering, A. (1984). *Constructing quarks: A sociological history of particle physics*. Edinburgh and Chicago: University of Chicago Press/Edinburgh University Press.

Pickering, A. (1986). Positivism/Holism/Constructivism. Paper presented at the weekly Colloquium of the Institute for Advanced Study, Princeton, NJ.

Pickering, A. (1987). Models in/of scientific practice. *Philosophy and social action*, **13**: 69-77.

Pickering, A. (1988). Understanding Alvarez: Articulating Kuhn. *Program, papers and abstracts for the Joint Conference*, British Society for the History of Science and History of Science Society, Manchester, England, 11-15 July 1988, pp. 133-8.

Pickering, A. (1989). Living in the material world: On realism and experimental practice. In *The uses of experiment: Studies of experimentation in the natural sciences*, ed. D. Gooding, T.J. Pinch & S. Schaffer, pp. 275-97. Cambridge: Cambridge University Press.

Pickering, A. (forthcoming). Big science as a form of life. To appear in *The restructuring of the physical sciences in Europe and the United States, 1945-1960*, ed. M. De Maria & M. Grilli. Singapore: World Scientific Publishing.

Pinch, T.J. (1985). Towards an analysis of scientific observation: The externality and evidential significance of observational reports in physics. *Social studies of science*, **15**: 3-36.

Quine, W.V.O. (1980). *From a logical point of view: Nine logico-philosophical essays*. 2nd ed. revised, Cambridge: Harvard University Press.

Rudwick, M.J.S. (1988). The closure of the great Devonian controversy. *Program, papers and abstracts for the Joint Conference*, British Society for the History of Science and History of Science Society, Manchester, England, 11-15 July 1988, pp. 155-9.

Shapin, S. (1982). History of science and its sociological reconstructions. *History of science*, **20**: 157-211.

Shapin, S. (1988). Following scientists around. *Social studies of science*, **18**: 533-550.

Stephanides, A. (1988). Concept extension and open-endedness: The case of quaternions. University of Illinois, mimeo.

Stevenson, M.L. (1987). The development of the hydrogen bubble chamber. In Trower (1987), pp. 105-7.

Suchman, L.A. (1987). *Plans and situated actions: The problem of human-machine communication*. Cambridge: Cambridge University Press.

Trower, W.P., ed. (1987). *Discovering Alvarez*. Chicago: University of Chicago Press.

Trower, W.P. (forthcoming). Luis Walter Alvarez (1911-1988). To appear in *The restructuring of the physical sciences in Europe and the United States, 1945-1960*, ed. M. De Maria & M. Grilli. Singapore: World Scientific Publishing.

Usher, A.P. (1954). *A history of mechanical inventions*. Revised ed., Cambridge: Harvard University Press.

Weiss, L. (forthcoming). The construction of CERN's first hydrogen bubble chambers. CERN History Project report CHS-26: to appear in *History of CERN*, vol. 2.

Woolgar, S. (1981a). Interests and explanation in the social study of science. *Social studies of science*, **11**: 365-94.

Woolgar, S. (1981b). Critique and criticism: Two readings of ethnomethodology. *Social studies of science*, **11**: 504-14.

University of Illinois

H.E. LE GRAND

IS A PICTURE WORTH A THOUSAND EXPERIMENTS?

1. INTRODUCTION

Experiment, in the traditional sense of some sort of carefully contrived and controlled interrogation of nature in a laboratory, has played in earth science a minor role for much of the past 150 years. Illustrations -- the maps, charts, photographs and drawings which form the 'visual language' of earth science and which fill its research literature and textbooks -- are commonplace. Rudwick (1976) tried to call attention to the neglect by historians, philosophers and sociologists of the visual element in science, particularly in geology, and traced the emergence of the visual language of maps and stratigraphic columns and sections in the 1830s and 1840s. Illustrations are ubiquitous in twentieth-century earth science literature. Much of the activity at earth science conferences and congresses centres on the aptly-named poster sessions. Rare is the lecture or paper which is unaccompanied by frequent use of slide and overhead projectors. These illustrations are not decoration, are not emblematic of a descriptive science, are not devices to break up blocks of texts, but are an integral part of discourse among earth scientists.

Pictorial representations are, of course, not unique to the earth sciences (see Latour, 1985; Lynch, 1985, 1987; Shapin, 1984). In what follows I shall discuss a few significant pictures of the earth from the 1950s and 1960s and then put forward suggestions as to the ways these functioned in the modern revolution in earth science associated with the theory of continental drift (see Wood, 1985; Le Grand, 1988). In the controversy over Drift resort was frequently made to pictorial arguments and evidence. Advocates of Drift, from its first full presentation by Alfred Wegener in 1915, especially made extensive use of a variety of images to advance their case and to undermine the orthodoxy of the fixed location of continents. I will suggest that in earth science illustrations serve some of the same ends that experiments serve in other sciences, that they were important in bringing about closure of the controversy over Drift, that they can be a powerful means of mobilizing

H. E. Le Grand (ed.), Experimental Inquiries, 241–270.

arguments in disputes, and, finally, that they -- like experiments -- make visible some presumed invisible order and effectively erase distinctions between 'representation' and 'reality'.

Latour (1985) suggests in a general discussion of illustration in science that scientists may be likened to explorers. To document their travels and to persuade their audiences of their veracity, they must return with 'material objects'. In experiments, the manipulation of chemicals or cultures or subjects is transformed by series of processes and inscription devices into images and text in a journal or monograph. The images -- graphs, curves, photographs, data tables and so on -- offer testimony, serve as witnesses of, the researcher's explorations. These 'material objects' should be portable, durable, capable of being shown to others, legible and combinable with other 'objects'. Ideally, they can be flattened onto pieces of paper and varied in scale. These objects -- these inscriptions and images -- can then be collected together in 'centres of calculation' (Latour, 1987: 232-47). There, with the aid of rulers and protractors or computers and statistics these 'pictures' of 'nature' can be transformed into sophisticated graphs complete with error bars or, even more economically, into equations; tied together in theories; and incorporated into texts so that the texts are not merely illustrated by the images but are a development and commentary on them, a unified discourse embracing the verbal and the visual. This characterization is readily applicable to pictures of the earth.

It is obvious how photographs might function in earth science. Provided we take as unproblematic the old adage 'the camera never lies', photographs give testimony that the objects under discussion do exist and do bear some resemblance to the account given. A description and definition of a 'syncline' (a downward warping or folding of strata) can be reinforced dramatically by a photograph of the 'real thing' and the reader not only is reassured that the author is talking about the 'real world' but also gains a mental template for identifying other synclines. I turn instead to other forms of illustration. Photographs *per se* seem to have played little part in the modern revolution, nor is this surprising given the obvious difficulties inherent in recording geological processes as opposed to the effects of those presumed processes. Most of the key research papers and first-generation textbooks made no use of them. Consider two extreme cases. Prior to the revolution, probably the most widely-used English-language text in physical geology (Holmes, 1944) contained 24 full-page and 173 half-page photographs of actual landforms and geological features. In the 1970s, just after the revolution, one of the most commonly used new texts (Wyllie, 1976) contained approximately 90 pages of diagrams, graphs, symbolic maps and cartoons but not one photograph. This change may have been emblematic of a shift from an emphasis on description to an emphasis on explanation, from 'the way the Earth looks' to 'the way the earth works', but clearly the visual element was important in both. 'Pictures' of data and 'pictures' of theories separately and superimposed were of crucial importance in the 1950s and 1960s (and indeed down to the present). Debate and discussion about Drift often centred upon whether or not these constructed pictures, these symbolic and abstracted images represented the Earth.[1] These illustrations were not merely illustrative but constitutive of the versions of Drift which eventually became the new orthodoxy.

2. MOTION PICTURES OF THE POLES

We begin with so-called polar-wander (P-W) curves which in the 1950s brought Drift out of the doldrums in which it had languished for over two decades. These images were acknowledged even by some opponents of Drift to provide impressive evidence for large-scale horizontal motion of continents.[2] Every step in the construction of these images was involved controversy and negotiation which often was focused upon images and resolved by the production of still more images. The end products -- the visually compelling P-W curves themselves -- are analogous to three-dimensional models of DNA in that they are the output of 'black boxes' compounded of technicians, data, theory, instruments, interpretation, and tacit knowledge. The construction of P-W curves and their deployment in the debate over Drift was an outgrowth of developments in the field of palaeomagnetism (studies of the earth's ancient magnetic field as recorded in igneous and sedimentary rocks).

Until the 1930s palaeomagnetic phenomena were of little interest. In 1938 researchers at at the Department of Terrestrial Magnetism (DTM) of the Carnegie Institution built a spinner magnetometer with which they were able to measure the magnetic orientation of sediments and clays. Their initial aim was to expand greatly the collection of observations on terrestrial magnetism made by the DTM as part of its charter to elucidate the origin of the earth's main field. Instead of relying upon 'real-time' observations of magnetic variation gathered by the DTM since 1905, measurement of 'fossil magnetization' might yield a record covering thousands or even millions of years. After the war, researchers at the DTM set new and more ambitious goals: to use rock magnetism to arbitrate between rival theories of geomagnetism and to address theories of large-scale crustal motion including Drift and polar wandering (see Le Grand, 1989). There were two rival theories of the origin of the geomagnetic field. An 'internal' origin had been postulated by W.M. Elsasser who suggested that geomagnetism arose from processes inside the Earth. E.C. Bullard developed a sophisticated version according to which geomagnetism was produced in a manner analogous to a self-exciting dynamo by the formation and circulation in the earth of thermal convection currents. The 'internal theory' was disputed by P.M.S. Blackett who in 1947 put forward a 'fundamental theory' according to which magnetic fields, including the earth's, are a fundamental property of all rotating matter. The DTM group thought the dispute could be resolved by palaeomagnetism: if the 'fundamental theory' be correct, ancient magnetic poles determined by palaeomagnetic studies should coincide with the present ones, assuming that the Earth's axis of rotation had not changed appreciably. John Graham, who had joined the DTM as a graduate student, carried out measurements on sedimentary rocks dating back approximately 200 MYA which seemed to indicate a general agreement between ancient and modern pole positions. He also proposed additional aims for palaeomagnetic studies. Some of these bore especially on geology; *e.g.*, their use in tectonics or in stratigraphic dating. One could also answer 'questions regarding large-scale movements of the crust, such as continental drift and polar wandering' (Graham, 1949: 160). Graham did not pursue this programme; instead, he shifted his attention to rock magnetism itself rather than its application to theories of geomagnetism or to Drift (Le Grand, 1989). However, the broader programme began in the early 1950s to bear fruit across the Atlantic. Several physicists, geophysicists and geologists in England in the 1950s pursued the global agenda of applying palaeomagnetism to geomagnetism, polar wandering and Drift.

The early focus of the English researchers, like that of the DTM group, was upon deciding between fundamental and internal theories of geomagnetism. Soon, however, their focus shifted to Drift. Blackett conceived a laboratory experiment to buttress his theory; an experiment which required the design and construction of a highly sensitive astatic magnetometer. His hopes were dashed: his results agreed instead with the views of Bullard and S.K. Runcorn. Blackett did not regard his negative result as fruitless nor did he give up hope for his theory; instead, he proposed a new use for his instrument. The astatic magnetometer was 'admirably suited to the measurement of very weakly magnetized geological specimens, for instance, certain sedimentary rocks'. He also called attention to the work of the Carnegie group, which he described as being of 'extreme interest' (Blackett, 1952: 312). A correspondence of ancient and modern magnetic poles would support his theory. A lack of correspondence would be evidence for internal theories: convection currents would presumably be unstable over geologic time and should give rise to apparent 'wandering' of the magnetic poles or, perhaps, non-dipolar fields (*i.e.*, several, simultaneous pairs of North and South magnetic poles). Blackett's group swiftly applied his instrument to palaeomagnetism. Meanwhile, Runcorn became involved in Jan Hospers' palaeomagnetic studies of Icelandic lava flows.[3] Like Blackett, Runcorn too thought that palaeomagnetic data, particularly possible dat on palaeosecular variation, might resolve the debate over the geomagnetic field. He launched his own programme of measuring rock magnetism.[4]

The Runcorn and Blackett groups soon began presenting their results in a more sophisticated manner than that of the the DTM workers. Graham and other early palaeomagnetists reported the results for each specimen in tabular form. Additionally, he and his colleaagues plotted the apparent pole of each specimen on an equal-area projection chart (Torreson *et al.*, 1949: 126). Little could be gleaned from such presentations of the data [See Figure 1]. One can easily ascertain

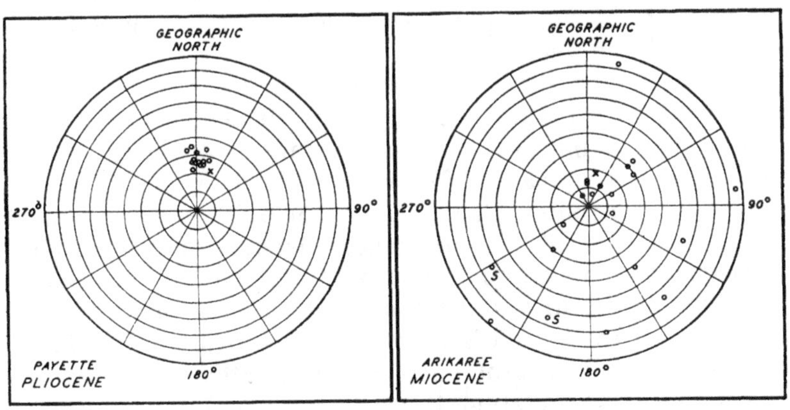

FIGURE 1a. Polarization in Rock Samples. Glenniserry, Idaho

FIGURE 1b. Polarization in Rock Samples. Badlands, South Dakota

X = Direction of Earth's Present Field
o = North Pole

(After Torreson, Murphy & Graham, 1949: 126)

whether the results for a formation were relatively well-grouped and therefore presumably consistent [Figure 1a] or widely scattered [Figure 1b] but there is no obvious criterion to rule out a datum as attributable to some form of error or poor sample quality. There seems to be no inherent theoretical overlay or structuring; rather, the raw data is presented with a minimum of filtering and processing. It is not even clear whether or not one should seek to collect samples which give well-grouped rather than dispersed readings. Graham's procedure of plotting every specimen was soon superseded by the application of statistics to yield average directions and to identify dubious data. Hospers had gathered readings on many specimens from both normally- and reversely-magnetized strata[5] and sought to reduce his data but found no suitable statistics applicable to a sphere were available. R.A. Fisher at Cambridge developed for him equations for expressing dispersion on a sphere which constituted a method of representing mean palaeomagnetic pole directions on the Earth's surface within a 95% circle of confidence (Fisher, 1953). The end result was for a given strata, which could perhaps be geologically dated, the mean direction of the geomagnetic field at that time. Fisher's methods were quickly adopted by Runcorn and, soon after, by Blackett's group. It was a simple step from there to calculate what became known as a 'palaeopole'. This was interpreted as representing the apparent position of the Earth's magnetic poles with respect to the strata when the strata was formed. Fisher's statistics and the calculation of palaeopoles were to be key components in the building of P-W curves.

The first P-W curve was constructed in the context of a dispute between the Blackett and Runcorn groups. Blackett's group, now based in London, had found by 1953 that contrary to previous studies there seemed to be a systematic difference between the modern poles and palaeopoles for Britain. These results could have been explained in terms of a wandering of the poles in accordance with an 'internal' theory of geomagnetism. His group instead proposed that England had rotated some 34^0 and moved toward the North Pole, an interpretation which did not run counter to Blackett's theory of the Earth's field (Clegg *et al.*, 1954). Runcorn and his group accepted Blackett's techniques and data but rejected his interpretation. Runcorn continued to favour an internal theory of geomagnetism and, until 1956, discounted Drift as an explanation of discrepancies between palaeopoles and the present magnetic poles. He proposed that the Earth's poles had moved with respect to land-masses, not that the land-masses had moved relative to one another. One of Runcorn's colleagues, in presenting their arguments, devised a new, suggestive and striking form of graphic representation which effectively fused arcane data, novel theory and a familiar Earth: P-W curves.

In the summer of 1954 Ken Creer, a physicist recruited by Runcorn, began writing up his thesis on the palaeomagnetism of the Old Red Sandstone and a series of other strata in Britain. Creer decided to display his results in pictorial form. He recalls (interview, t1 s1) that he was inspired to use this form of representation by an illustration in a geophysics text.[6] That illustration depicted the supposed wandering of the geographical north pole, as assessed through palaeoclimatic evidence, superimposed on a map of the Earth. Creer plotted his palaeopoles in a similar fashion, in part to see if the palaeomagnetic and palaeoclimatic data were consistent.[7] He connected them in what constituted the first palaeomagnetic P-W curve [See Figure 2].

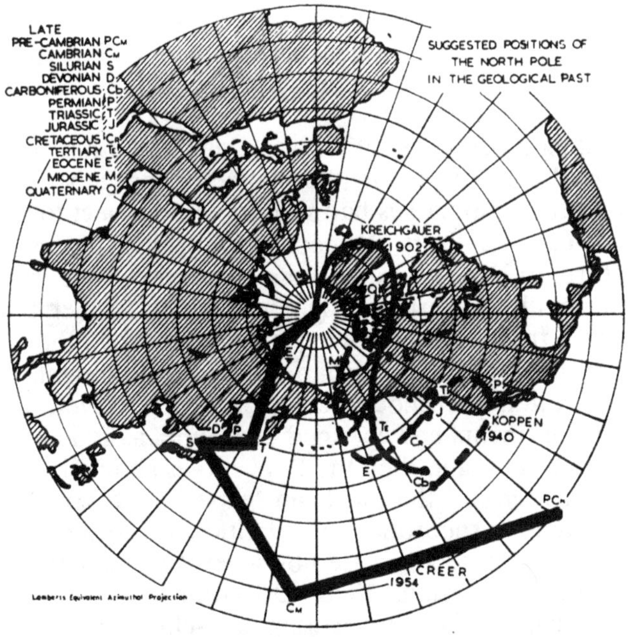

FIGURE 2. Creer's P-W Curve (From Creer, Irving & Runcorn, 1954: 165)

This was unveiled at a meeting of the British Association and created something of a sensation. It was the form of representation, rather than simply Creer's claim that his results indicated large-scale crustal movement,[8] which captured the attention of others. Blackett's group at that same meeting reported in tabular form similar results and represented them in terms of a former location of Britain and supported a more definite pro-Drift interpretation.[9] But, it was Creer and his P-W curve which featured not only in ensuing scientific discussions but especially in the popular media: *Time* magazine, the *Times*, *Guardian*, *Telegraph* and other newspapers from Bangkok to Arizona. At a less obvious level, Creer's image in a sense brings the past into the present and blurs a distinction between earth science considered as a historical science or as a physical science. It also brings the spatially distant first into a small corner of England and then into the compass of a single page. Rocks and strata from hundreds of kilometres and millions of years distant are al assembled into one compact, easily grasped 'picture'. Creer's (and later) P-W curves are in appearance simple, yet are set in the context of a web of theoretical and technical assumptions, are based upon meticulous fieldwork, reflect hundreds or thousands of discrete and time-consuming measurements and incorporate fairly sophisticated statistics. Other palaeomagnetists quickly adopted this form of representation.

Debate among palaeomagnetists arose as to whether the crust as a whole had moved with respect to the magnetic and geographical poles or Britain and other landmasses had moved with respect to one another and to the poles; *i.e.*, Drift. Runcorn and some of his colleagues favoured the former; Blackett and others, the latter. Both groups agreed that the matter could be resolved by constructing further pictures; *i.e.*, by gathering data from and producing P-W curves for other continents. If the continents had moved as a group, as Runcorn thought, then curves from different continents should coincide. If, however, the relative positions of the continents had changed over time, as Blackett and other proponents of Drift thought, then the P-W curves should be different. In short, b th groups agreed that such an 'experiment' could distinguish between two models or theoretical possibilities. Illustrations in this case had a role similar to a role advocated by several philosophers of science (Giere, 1988: 165-7; Laudan, 1978; Donovan *et al.* 1988: 38-9 and *passim.*) for experiment. Somewhat tongue-in-cheek, we could think of these as 'Crucial Illustrations' instead of 'Crucial Experiments'.

Runcorn switched his position in 1956 and accepted Blackett's view that Drift was needed. Pictures were crucial in his decision. In a paper published early that year, he had reaffirmed his opinion that the discrepancies between polar positions given by American rocks and European (including British) rocks were minor and that both sets of data could be explained by a relative movement between the poles and the crust without drift. However, in responding to counter-arguments from his colleagues and on re-evaluating his data, he soon rejected his earlier interpretation. The north magnetic poles calculated from North American data seemed to lie systematically to the west of those calculated from European data. Moreover, the two curves had converged until they intersected for modern times at the present magnetic pole. Given his assumptions that the Earth at any one time had only one set of magnetic poles and that these were roughly aligned with the poles of rotation, the only way he now believed he could match up the P-W curves was by postulating that North America had been displaced since the Triassic (about 213 MYA); *i.e.*, the *convergence* of the curves since the Triassic represented a *divergence* of the continents. This, he saw, provided a powerful argument for Drift and he warned that 'the geophysical objections to continental drift' (objections which he had himself endorsed only the previous year) 'must be re-examined' (1956: 253). As P-W curves from other continents were mapped, further weight seemed to be added to a Drift interpretation [See Figure 3].

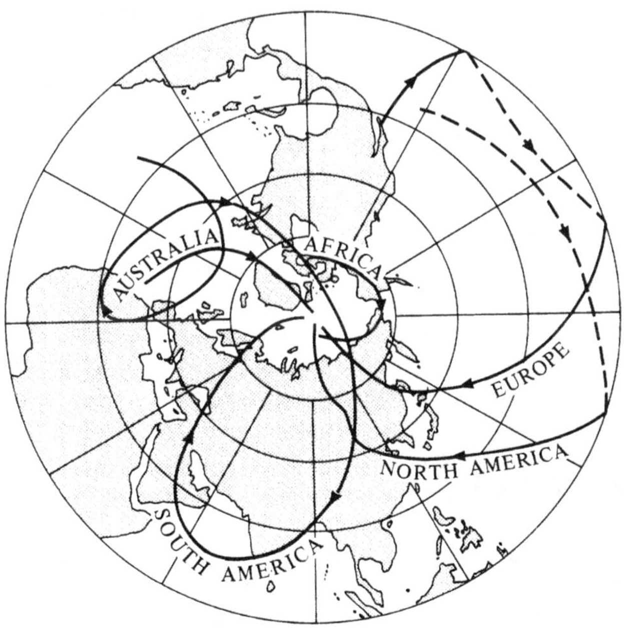

FIGURE 3. Schematic P-W Curves for Different Continents (After Runcorn, 1962: 24). Note especially the North American and European curves.

Without Drift, the assembled curves seemed quite chaotic to indicate multiple sets of randomly moving magnetic poles (or a crust with the strange property of moving in several directions at the same time); with Drift, all the different curves could be harmonized in terms of one set of more-or-less stable poles and wandering continents.

From the mid-1950s Drift became central to the research agendas of both the Blackett and Runcorn groups and, for this as we have seen, pictures were in no small measure responsible. These pictures were the product of many individuals: physicists, geologists, geophysicists, technicians and support staff. By the early 1960s most palaeomagnetists agreed that the divergent paths from different continents could be reconciled only through Drift. By then, for those who constructed these curves it appears that the acid test of authenticity had come to be whether or not the images accorded with some form of Drift theory. If a researcher found a set of palaeopoles which coincided with the modern poles, this was taken as *prima facie* evidence of 'overprinting' of the true ancient magnetism[10] rather than evidence of a congruence of ancient and modern poles as Graham had thought. A theoretical picture served as a test of the quality of data and, in this sense, the methods of production of that data were attested to and guaranteed by the end product: newly-calculated palaeopoles were expected to fit at least roughly the P-W curves which had previously been constructed. The relationship between P-W curves and Drift was a symbiotic one. If Drift were rejected, then the patterns -- the divergent P-W paths -- elicited from the scatter of apparent ancient pole-positions were meaningless. Perhaps researchers would have turned to other phenomena or perhaps this field of investigation would have reverted to a marginal field of little consequence. Drift also served a heuristic role; *i.e.*, it suggested ways in which Runcorn, Blackett and others could develop their agenda: the interpretation of diverging curves in terms of relative movement of landmasses; Runcorn's work on convection currents as a mechanism for Drift; at an empirical level, the employment of palaeoclimatic data as a cross-check on and reinforcement of palaeomagnetic data; at a 'technique' level, guidelines as to appropriate locations and approximate dates of deposits which might be studied. Conversely, palaeomagnetists provided fresh argument for Drift.

It is a moot question whether or not palaeomagnetics and P-W curves alone or even in concert with palaeoclimatology and palaeobiogeography would have been sufficient to lead to widespread acceptance of Drift. The line of research pioneered and sustained by Blackett, Runcorn and their associates sparked renewed interest in Drift which transcended the specialty in which they worked and gave the impression that Drift was progressing empirically and theoretically.[11] Even if there were many possible objections to palaeomagnetists' data, techniques, methods, instruments and results, even if the same could be said of the representation of results as P-W curves and the interpretation of these curves as evidence for Drift, these curves and the data on which they were based were taken seriously. The data were not regarded as artifacts; they had not been conjured up out of thin air but out of solid rocks. By the early 1960s reproductions and discussions of these pictures indicates that a stabilization of the theory, data, techniques and instrumentation involved had occurred. However, other kinds of pictures, pictures of the inaccessible and the invisible, proved to be more crucial than P-W curves in the modern revolution.

3. ILLUSTRATING THE INVISIBLE

In the 1960s other illustrations and images were added to the arsenal of Drifters. Those that seem to have been the most powerful in winning the allegiance of others were associated with the seafloor-spreading version of Drift (see Frankel, 1979, 1980, 1982; Le Grand, 1988: 170-82 & 196-218). A common objection to earlier versions of Drift had been that it was physically impossible for the continents to have 'drifted' through the solid rock of the ocean floors in the manner Wegener had envisaged. In the seafloor-spreading version, adumbrated by Harry Hess, the continents rather than moving *through* rigid crust were embedded in and moved along *with* that crust. New ocean floor material was created along the network of mid-oceanic ridges; the older seafloor -- and the continents riding on that crust -- gradually moved away from the ridges; and, completing the model, older seafloor was consumed in the grat ocean deeps, the trenches. Most of its early proponents were specialists in marine geology and geophysics and most of the more significant evidence and arguments in its favour were drawn from the three-quarters of the Earth's crust -- the oceans -- inaccessible except through remote-sensing devices. It was only in the late 1960s and early 1970s that this version began to be applied to problems of continent geology. Images of theory and images of the earth pervaded these developments and show the power of illustrations not only to bring the past into the present but also the inaccessible into the laboratory, the study and the lecture hall.

The most persuasive evidence for seafloor-spreading versions of Drift came from magnetic studies of the ocean floors. The material composing the oceanic crust had been thought to be relatively undifferentiated. But, when measured with magnetometers towed from the surface, there were surprising variations in the strength of the magnetic field: regions with a higher-than average field alternated with regions of a lower-than-average field. Beginning in the late 1950s marine scientists systematically measured and plotted these magnetic anomalies. The results were portrayed as 'maps' of seafloor-striping [See Figure 4].

FIGURE 4. Seafloor 'Striping': Juan de Fuca off Vancouver (After Raff & Mason, 1961: 1268, by permission of the Geological Society of America)

Such zebra-like patterns were generated by representing positive anomalies (strong fields) as darker sections and negative anomalies (weak fields) as lighter sections on maps of the floors. A variety of explanations of the origin of these patterns were put forward in the late 1950s and early 1960s (*e.g.*, that the stripes reflected differences in chemical composition and associated susceptibility to magnetization) but none generated much enthusiasm.

In the mid-1960s Fred Vine and Drum Matthews proposed a version of seafloor spreading which incorporated magnetic striping. This is readily visualizable in cartoon form [See Figure 5].

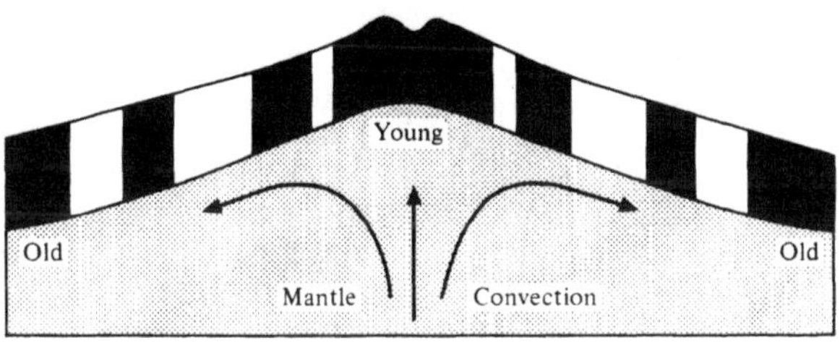

FIGURE 5. Simple Model of Spreading & Magnetization

Molten material continually wells up at the ridges and, as it cools, acquires a magnetization in accord with the Earth's magnetic field at that time. Alternating bands of normally- and reversely-magnetized seafloor result from reversals of the Earth's magnetic poles. When measured with a towed magnetometer the magnetization of those strip(e)s of normally-magnetized seafloor is 'added to' the Earth's present field to give a 'positive' anomaly; for reversed strip(e)s, it is 'subtracted from' the Earth's present field to give a 'negative' anomaly. The process by which the strip(e)s are formed is a continuous one so that in effect the seafloor is a 'tape recorder' of alternating periods of normal and reversed geomagnetism.

This proposal initially excited little interest. The key assumptions of polar reversals and seafloor spreading were thought improbable. Moreover, Vine and Matthews' explanation of striping appeared *ad hoc*: it was ingenious but there seemed to be no evidence which favoured it over alternatives. In 1965, however, Vine in association with Tuzo Wilson worked out an implication which should be evident in pictures. If the new seafloor were created symmetrically along the axis of the ridge and moved away from either side of the ridge at the same rate, the pattern of striping on one side of a ridge should be the mirror image of the pattern on the other side. A magnetometer towed by a ship perpendicular to a ridge axis should record a 'profile' of 'peaks' (positive anomalies) and 'valleys' (negative anomalies) with a bilateral symmetry along the ridge axis. Vine and Wilson began inspecting maps of seafloor striping. They were able to identify some such symmetries; for example, on the Juan de Fuca Ridge, and such evidence won increasing support in the small community of marine geologists and geophysicists. The arguments and evidence were visual: pictures which juxtaposed theory with anomaly maps and profiles in a striking fashion [See Figure 6].

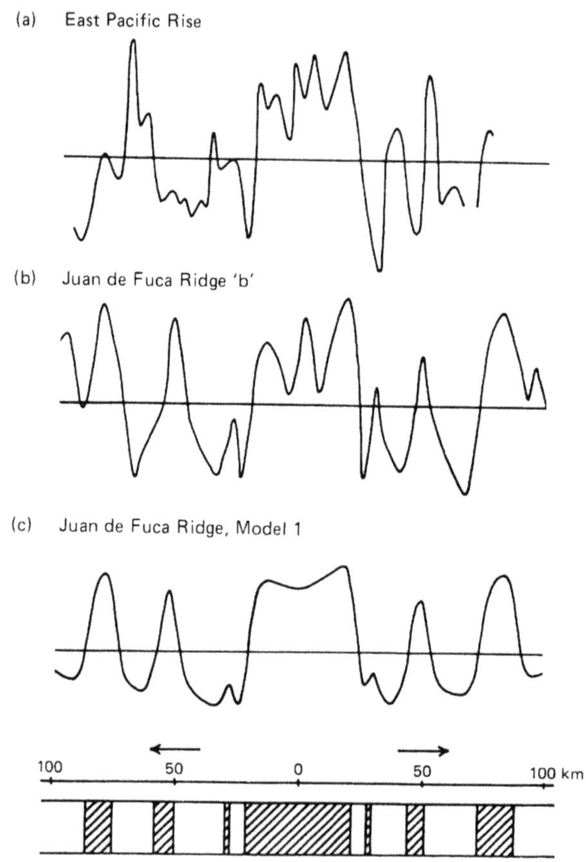

FIGURE 6. Juan de Fuca (after Vine & Wilson, 1965: 487)

In Figure 6 theory and data are presented simultaneously. The 'data'; *i.e.*, the 'profile' or processed representation of magnetometer readings across the ridge crest ('0' on the distance scale at bottom), are depicted as *Juan de Fuca Ridge 'b'*. A schematic version of the theory (compare with Figure 5) is drawn at the bottom. In the middle is an imagined (seen with the mind's eye but captured on paper) representation of an ideal profile corresponding to the theoretical model. Viewers would, as intended, be struck by the 'close' approximation of the visual data to the visual theory: the evidence is literally right in front of their eyes. A later variant [See Figure 7] exemplifies even better the juxtaposition of theory with data and the way in which images can make the invisible and inaccessible both visible and immediate.

In Figure 7 is presented not only a highly schematized theory (bottom, with age and distance scales) but also a very concretized one: a representation of an 'actual' cross-section of the 'real' oceanic crust which has in one sense been 'seen' by no one since it is several thousand metres beneath the surface and probably partially buried under accumulations of sediments but which is now here to be 'seen' by anyone. Moreover, here are presented not only the 'real' and 'ideal' profiles but also a section of the 'real' map of anomaly patterns, though of course the stripes could be 'seen' only through magnetometers and computers. Even with the production of such images the Vine-Matthews-Wilson version met resistance. Such illustrations offered a point of attack. In many other profiles and anomaly maps the symmetry across the ridge axis was not well-marked. The degree of symmetry, the 'closeness' to a mirror image may have lain in the theoretical eye of the beholder. Researchers at the Lamont-Doherty Geological Observatory, who had accumulated an enormous stock of such images, and some of whom preferred non-Drift explanations of the patterns, remained skeptical -- until they themselves produced such compelling images of symmetry as those across the East Pacific Rise and the Reykjanes Ridge.

The clinching argument for most marine geologists and geophysicists was the single profile collected for Lamont by the research ship *Eltanin* on its nineteenth traverse across the East Pacific Rise [See Figure 8]. Other profiles made on the same voyage were set aside in favour of *Eltanin*-19. One could literally fold the profile along the ridge and match up anomalies from the opposite sides.[12] It became the template against which other profiles were matched, evaluated and resolved. To see this visual 'proof' was to become a believer in the symmetry thesis, in Vine-Matthews, and in what soon became known as plate tectonics. Vine himself commented when he first saw this profile, 'It was all over but

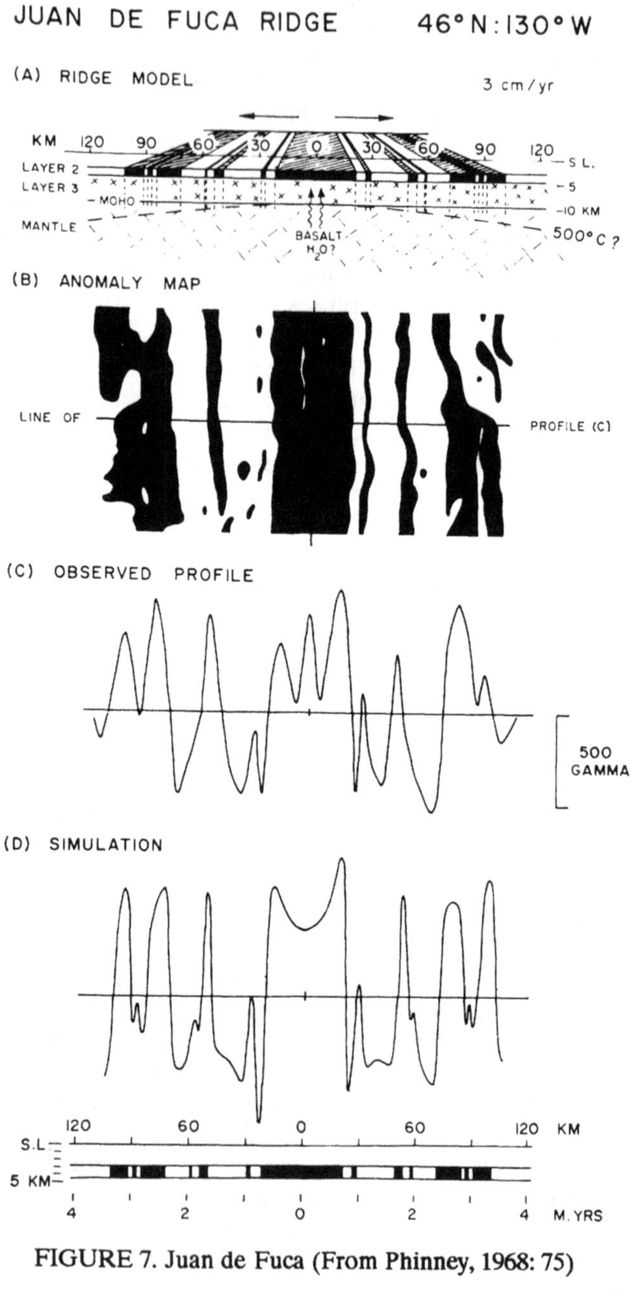

FIGURE 7. Juan de Fuca (From Phinney, 1968: 75)

the shouting'. The reaction of marine geophysicists and palaeomagnetists to this single image were similar: N. Opdyke: 'That's it -- you've got it!'; R. Doell: 'It's so good it can't possibly be true, but it is'; A. Cox: '...a truly extraordinary experience...there was just no question any more that the seafloor-spreading idea was right'; W. Pitman: 'It hit me like a hammer'; or, even more tellingly from J. Worzel, a staunch opponent of Drift: 'It's too perfect' (quoted in Glen, 1982: 333-9; Menard, 1986: 264). By the late 1960s the revolution was essentially concluded in the specialties of palaeomagnetism and marine geology and was spreading rapidly to other areas of earth science.

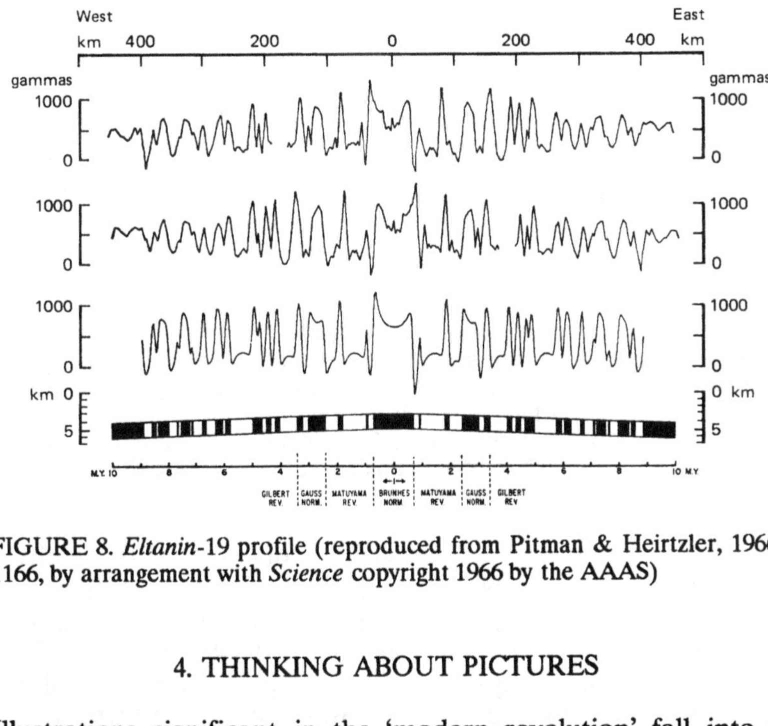

FIGURE 8. *Eltanin*-19 profile (reproduced from Pitman & Heirtzler, 1966: 1166, by arrangement with *Science* copyright 1966 by the AAAS)

4. THINKING ABOUT PICTURES

Illustrations significant in the 'modern revolution' fall into three broad categories. First, there are *symbolic maps*: pictures of such data as apparent polar wander, magnetic anomalies, ocean floor topography, and earthquake centres which fuse the familiar with the new. The skeletons of these 'data pictures' are recognizably the Earth but an Earth reduced to a schematic outline of continents and in which attention is focused upon structures and phenomena and data accessible only through instruments. A second category would be *symbolized theories*; *e.g.*, pictures or diagrams of seafloor-spreading or plate motions.

Figure 9 is one of the most frequently-reproduced diagrams of this type. Such 'theory pictures' depict geological processes in schematic, almost cartoon form. It deftly summarizes the essentials of the plate tectonics version of Drift: transform faults, trenches, ridges, plate collisions, *etc.* Third, there are many illustrations in which symbolic maps and symbolized theories are combined; *e.g.*, Figures 6 and 7.

The developments we have discussed could be glossed in various, more familiar ways as promoting the closure of controversy over Drift. Kuhn, L. Laudan and others who emphasize the importance of problem-solving in science might point out that for Blackett, Runcorn and others

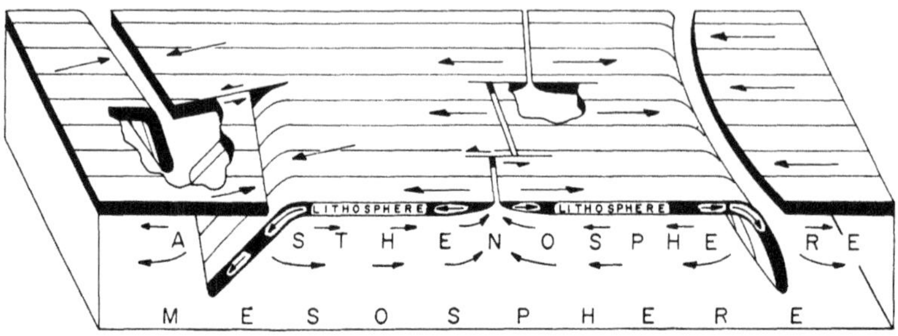

FIGURE 9. Plate tectonics as symbolized theory (after Isacks, Oliver & Sykes, 1968: 5857). The plates form the rigid lithosphere. New plate material is formed at a ridge (centre) which is offset by two transform faults. A simple convergence of plates (right) results in the subduction and consumption of old plate material in a trench. A complex pattern of convergence (left) results in two trenches joined by a transform fault.

Drift solved the problem of reconciling divergent P-W paths from different continents, a problem which did not arise until four decades after Drift had first been proposed. This would be in accord with the suggestion (*e.g.*, Kuhn, 1977: 322; Lakatos, 1978: 49, 52, 69, 112; Laudan, 1984: 100) that the ability of a research programme to solve new, unanticipated problems -- problems it was not invented to solve -- is a powerful reason to prefer that programme. In Laudan's terminology, palaeomagnetists thereby created for Drift a whole new class of solved problems; for its rivals, a new class of empirical anomalies. Additionally, the linking of palaeomagnetics with other specialties may have given greater weight -- or at least new visibility -- to some of the older anomalies facing Drift's rivals; *e.g.*, the great southern glaciation (See Frankel, 1984). For historians and philosophers who subscribe to the notion of crucial experiments, what may be especially relevant to the possible role in this case of pictures (or experiments) arbitrating among rival theories is that there was no logical dependence of the 'experiment' on one or other of the rival programs. The existence of rock magnetism, its detection and measurement, the plotting of apparent ancient poles and the construction of polar-wandering curves could be performed without necessary reference to one or another of the rival theories of the earth (see, *e.g.*, Giere, 1988: 138-9; Donovan *et al.*, 1988: *passim.*). This is certainly not to deny that palaeomagnetics was 'theory-laden'; rather that the 'ladenness' was not owing to what were considered to be the theories under test. A common theme in philosophy of science is the importance of successful predictions (*e.g.*, Lakatos, 1978: 16-17; Laudan, 1977: 43; Giere, 198: 198-203). This offers yet another gloss of the closure of controversy in favour of Drift, and one to which some palaeomagnetists themselves subscribed. The construction of P-W curves which could be fitted together through Drift was regarded as a confirmation of Drift and, inferentially, of collateral assumptions of about the stability of magnetism, the unproblematic character of reversals, the axial dipole view of the Earth's magnetic field and so on. This view was not dismissed out-of-hand even by those less sympathetic to Drift. Similar claims can be mounted with respect to seafloor-spreading, Vine-Matthews and images of the ocean floor, especially the *Eltanin*-19 profile. Irrespective of such familiar philosophical or methodological glosses -- and the problems associated with them -- the pictorial dimension involved in these developments was of immense importance in its own right.

For proponents of Drift, the three types of pictures constituted a powerful weapon to achieve closure on their terms.[13] Pictures of data and processes functioned in a variety of ways. At a simple level symbolic maps summarized vast amounts of data in forms which were concise, manipulable, easily learned and easily remembered. At a glance viewers -- once their eyes had minor training -- could assimilate the content of innumerable research reports and papers, thousands even hundreds of thousands of individual bits of data. Symbolized theories summarized complex geological processes. Such images constituted painlessly consumable distillations of assumptions, arguments, equations. Confronted with these representations of theory, the viewer did not have to read vast quantities of research literature or learn arcane physics and mathematics in order to grasp the key points of a theoretical model and to make some use of it.

These summaries could be juxtaposed or combined with or superimposed on one another. Images of oceanic ridges and trenches as 'seen' in Heezen and Tharp's imaginative topographical portrait of the sea floor, patterns of seafloor-striping and plots of earthquake centres could be merged and superimposed on, and indeed help define and constitute such images of theory as tectonic plates and transform faults. These taken together yielded a unified geophysicist's earth exemplified in the plethora of illustrations of the 'new global tectonics' ranging from those in *National Geographic* and *Scientific American* to most of the recent earth science textbooks as well as in maps, posters, wall charts, computer simulations and videos. Conventional distinctions between the 'natural' and the 'artificial', between 'nature' and 'laboratory' are blurred by plotting these inscriptions and images upon familiar images of the 'real' world. Similarly, recall that P-W curves were not simple line graphs, they were plotted on the surface of the 'real' globe and were readily associated in the viewer's mind with theories of crustal mobility. Indeed, Runcorn and his colleagues began to make extensive use in their research of very large models of the Earth on which cut-outs of the continents could be moved about and onto which P-W curves could be plotted.[14] In brief, these images of data and theories, plotted on the Earth, were seductive mnemonic devices.[15]

These pictures were instrumental in the dissemination and the concentration of knowledge and the mobilization of resources. I have discussed elsewhere (*e.g.*, 1988; 1989) the importance of differential access to techniques and information in influencing the attitudes of geologists to competing global theories and how specialist technical and social interests could shape positive or negative attitudes to Drift and its rivals. In proposing, defending or evaluating Drift, symbolic maps and symbolized theories provided pre-digested theory, argument and evidence. Professional geologists (and students) for many purposes need only see and remember a series of images. This may have had special significance in Britain, America and Australia in which the boundaries among specialties, seem to have been more sharply defined and difficult to cross than elsewhere. Academic departments were, for example, typically staffed with one or more specialists in each major division but with no one formally responsible in their research or teaching for an integrated picture of the whole Earth. The traditional core of earth science was the Earth of stratigraphy and palaeontology. Geophysics and marine geology were very much on the fringe. Much of the new evidence, the new theories and the new recruits for Drift came from that fringe in the 1950s and 1960s. Drifters could concentrate, could draw together in a coherent, persuasive, compact whole the new images of theory and data from their own specialties and add older images from others (*e.g.*, maps of the jigsaw fit of continents, palaeoclimatic and biogeographical maps). These could then be disseminated among other specialties and the general geological community. Such pictures could thereby mobilize theory and data across specialty boundaries and be used to enlist allies from a range of specialties. This may help explain how knowledge of the new versions of Drift could spread rapidly over diverse specialties in the late 1960s and early 1970s and, perhaps also, why many of those who adopted Drift in those years had not and did not read many of the key papers.

These characteristics of pictures have obvious relevance to pedagogy. A few diagrams, a few illustrations sufficed to convey the essentials of the new orthodoxy. Images could be transported directly from the research literature into general textbooks. The core sections of the post-revolutionary textbooks were those on Drift and these reproduced many of the images we have encountered. These abstracted, constructed images reflect the increased social and technical power and centrality of geophysics within the discipline. Palaeontology and stratigraphy -- usually still illustrated by photographs which captured the messy details of the Earth -- were often relegated to the fringe. In textbooks as in the research literature, symbolic maps and symbolized theories have become exemplars of the methods of the new geology and emblematic of its orientation toward physics and away from natural history.

Pictures also served heuristic functions. Symbolic maps of earthquakes, volcanic activity, ridges, trenches and faults led Tuzo Wilson to his 1959 hypothesis of a world-wide fracture system and thence to his 1965 model of a crust composed of rigid plates, mobile belts and transform faults -- a model subsequently transformed into a symbolized theory. S. Warren Carey's studies of tectonic maps led him to the global theory of Earth Expansion. Or, consider P-W curves. Scientists could not store, interpret, compare and integrate all of the discrete pieces of data in their heads. One might be able to make sense of a small number of individual readings but the larger patterns were not evident until these readings were combined, massaged and 'mapped out' as a palaeopole. One might hold in one's mind several such poles but again the larger pattern would not be evident until these poles were mapped out as a P-W curve. Finally, one might be able easily to 'visualize' in one's mind a particular P-W curve and perhaps recognize that some individual measurements from another continent or even its calculated palaeopoles did or did not roughly coincide, but one would unlikely be able to recognize apparently systematic divergences or the feasibility of reconciling such divergences through the drift and rotation of now separate landmasses until these divergent paths were mapped out.

Finally and more speculatively, pictures not merely represented but constituted the real world for many younger geophysicists and some geologists who matured during the 'revolution'. Fred Vine's thesis problem focused not on the sea floor itself, but upon multiply-processed images of what presumably lay beneath the water. The arguments and evidence for his theory were drawn from paper: they were only indirectly grounded in the solid rock of the Earth. For Dan McKenzie, the first across the finish line in the race to quantify Wilson's plate tectonics, the Earth *was* the paper earth of symbolic maps and symbolized theory bound into journals and monographs. McKenzie's plates were stripped of rocks and valleys and strata, stripped of all the wealth of detail patiently gathered by legions of field geologists. The problems to be solved, the methods to be used, the data, the solutions proposed were shaped by pictorial inscriptions and images. McKenzie's mathematical skills stood him in good stead here for those pictures enabled and often represented the mathematization of processes and mathematical renderings of data. Perhaps one of the seductive aspects of plate tectonics was that it combined powerful visual elements with a an element of 'exactness' common in physics or chemistry but previously far less common in geology.

I return to my starting point: the relative absence of experiment and ubiquity of illustration in the earth science teaching and research literature. As we have seen, illustrations occasionally filled some of the roles traditionally assigned to experiment. In particular, images could serve as representations of natural processes, 'confirmations' of a theory, demonstrations of a theory for consumption by other researches or for pedagogical purposes and as exemplars. Like experiments, in which nature is brought into the laboratory and domesticated, images were often constructed abstractions of natural processes in which non-essential details were suppressed or omitted. Illustrations could, perhaps, in some circumstances serve as 'crucial experiments' and also play a heuristic role. Let us explore some less obvious similarities between illustrations and experiments.

First, illustrations are theory-laden. We might think of analogies between experimental facts and illustrational facts. The data and pictures constructed from them were given significance by Drift; the illustrations, especially those combining symbolic maps and symbolized theory, conferred plausibility on Drift. In the case of P-W curves, if Drift were rejected, then the disparities among curves from different continents might have been 'noise', the legitimacy of the set of techniques might have been called into question and perhaps the results would have been set aside as artifactual (in the pejorative sense). In the context of Drift, however, these curves were seen as highly significant new evidence, perhaps even the palaeomagical path out of the impasse between Drift and its rivals which had existed since the 1930s.[16] As a coda to this point it is not far-fetched to think of many of these illustrations as counterparts of the experimentalists' black boxes. For P-W curves, each link in the chain from strata to relative motion of continents was constructed through a process involving the resolution of theoretic-empirical disagreement and debate. But, in the end-product, those traces of construction and disagreement were erased. Presented with the final images, a non-specialist would have faced a daunting task to dismantle them individually or ensemble in the 1960s and 1970s. A 'doubting Thomas' who wished to put forward a rival interpretation would have had to unpack, to re-open those black boxes to retrace their construction -- an undertaking requiring enormous expenditure of resources.

Second, illustrations like experiments may render visible some presumed invisible order. Perhaps the plainest example of this which we have encountered is seafloor striping and its explanation by Vine, Matthews and Wilson. I offer as anecdotal evidence of the transforming of the invisible into the visible the belief occasionally encountered among students that if they could swim down to the ocean floor off Vancouver and sweep away the sediment, they would 'see' a striped seafloor. The illustrations and the language of the associated texts do reinforce that natural impression. At a more sophisticated level, the *order* in that pattern became *visible* through the addition of a symbolized theory. Connected with this point is the role of illustrations in what Shapin (1984) has termed 'virtual witnessing'. There is the seemingly trivial question of whether the spinner and astatic magnetometers were measuring something that pre-existed or were creating the phenomena they were supposedly detecting. The question is not an absurd one. A Kuhnian (1970: 24 & 192-4) might claim that palaeomagnetism of sedimentary rocks might not have been investigated and the necessary instruments invented but for the existence of theories in physics, geophysics and geology which directed attention to them. At a more subtle level one might claim that without these instruments and associated techniques, ancient polar positions and P-W curves would not exist for scientists: 'the phenomena are thoroughly constituted by the material setting of the laboratory' (Latour & Woolgar, 1979: 64; Latour,

1987), they are not given to us by nature.[17] For researchers who did not have access to the necessary equipment, the phenomena in an important sense did not exist -- whether those researchers opposed or supported the agenda and interpretations. Those scientists who did have access may have thereby gained expertise, authority and even immunity from some types of criticism but they faced a difficult task in persuading others of the well-foundedness of their interpretations especially if they went against firmly held beliefs. Pictures of data in particular, embodied in text presenting those interpretations, effectively said: 'It is alright if you don't have the equipment and expertise. If you join us in our lab, here is the data you would get and it is as plain as the nose o your face that it supports our interpretation.'

I have in my argument begged many questions. Two are particularly salient and warrant a preliminary response. The first is in the following vein: What is the relevance of the distinctions often drawn between observation and experiment? I would reply, 'not much'. A distinction based on data obtainable by our unaided senses and data obtainable only through instruments seems untenable. After all, as Latour (this volume) suggests, we could -- somewhat tongue in cheek -- classify a scientist equipped with pencil, pen or word processor as an 'inscription device'. Some distinctions could perhaps be defended; *e.g.*, observation as a passive activity in which no special effort is made to control variables. Clearly a procedure involving the running of many trials in a laboratory while systematically altering the variables has limited application in earth science: the results of the more interesting experiments would not be known for a very, very long time! Nonetheless, it is clear that the construction of many of the images we have considered is far from a passive recording of passive observations. Consider the construction of P-W curves. This process involved a definite intention; *i.e.*, initially to test theories of geomagnetism and later, to 'prove' or 'disprove' Drift. It involved as well the design and use of special instruments, the development and application of mathematical tools and the development and use of various 'quality control' procedures: *e.g.*, sampling techniques, avoiding metamormophosed and weathered deposits and the development of 'cleaning' apparatus and techniques. Further, it involved the use of simple experiments both to develop 'observational' techniques and to underpin assumptions; *e.g.*, several researchers experimented with grinding up samples and allowing them to precipitate in the Earth's field to see if alignment with that field occurred. What line there might be between the sort of 'observation' employed in P-W curves and 'experimentation' seems a blurred one indeed (see Hacking, 1983: 173-4, 186-209 & *passim.*; Giere, 1988: 125-8).

The second question touches on the special character of visual materials: Is the characterization I offer of illustration generalizable to fields of science other than geology and geophysics? My response here is more tentative. The relative quantity of and emphasis on illustrations may well vary among different fields of science but the multifarious roles of the visual element in earth science are not unique to it. At the very least, in descriptive sciences or theoretical ones, in field sciences or laboratory ones, illustrations can mobilize a wide range of theoretical and empirical arguments including experimental results and make them appear 'natural'. In opening, sustaining and closing scientific controversies, how 'facts' are marshalled and presented seems, at least on occasion and especially for the non-specialist, to be crucial. Pictures, images, illustrations appear to be extraordinarily effective in this connection. Why is this the case? Obvious answers have surfaced in the course of our study thus far: illustrations do have the characteristics of Latour's 'material objects', illustrations summarize (as do equations) enormous amounts of detailed information, illustrations can serve many of the functions commonly ascribed to 'experiments' in a traditional sense.

Recent work in the loosely-knit field of cognitive science suggests other, fundamental reasons for the power and importance of pictures of theory and data. Two uncontroversial findings reinforce reasons assigned above for the importance of illustrations. First, with respect to mnemonic functions, it seems that pictures which combine, in a relationally organized fashion, either several symbolic maps (diverse data) or symbolic maps with symbolized theory are more effective if 'inputted' as composite pictures than as separate ones. Second, maps even in schematic form seem to assist markedly the assimilation and retrieval of textual information contained in geographical texts. This, of course, corresponds to the standard practice in earth science lectures, papers, 'posters' and textbooks. More speculatively, it is argued by many researchers that humans are 'hard-wired' in such a fashion that our visual and verbal processing systems function independently in the encoding, storage, manipulation and use of information. If so, there are special types of representation and processes for dealing with spatial or visual information (pictures) as distinct from a propositional characterization of verbal and linguistic information (texts). If this be the case, then pictures of theory and data are important not only for the conventional reasons mentioned but also because they are assimilated in a special way by the brain and, moreover, may require far less processing for storage, retrieval and perhaps integration with propositional information. There is more general assent in the field to the importance of 'cognitive maps' and 'mental images' ('pictures' in the mind) and 'mental models'. This also has bearing on our response. Mental models may be thought of as representations (theories) of aspects of the external world which can be analyzed, cri icized and revised in terms of similarities in both kind and degree between the model and the external world. In this view, mental images are a special class of mental models in which the processes involved in the experience of a mental image (the 'seeable' counterparts of the model in the external world) are similar to those involved in the experience of an external picture or object. If this be the case, then pictures of theory and data provide pre-processed and predigested 'maps', 'images' and 'models'. Is a picture worth a thousand words? The answer given by cognitive science is an unqualified yes.[18] Is a picture worth a thousand experiments? The answer from our examination of a few facets of the modern revolution in earth science is yes.

NOTES

1. Throughout, I use the term 'Earth' to indicate that which is studied by field geologists, traversed by explorers and represented in photographs, geographical maps, relief maps, atlases, globes, and so on. The terms 'earth' is used for the geophysicist's earth which is presumed to have some resemblance to the Earth but in a still more indirect, artificial and constructed manner.
2. For more general histories of palaeomagnetism, polar-wander and Drift, see Frankel (1987), Irving (1987) and Le Grand (1988: 138-69). Glen (1982) provides a detailed history of young-rock dating, polarity reversal timescales, and several aspects of marine magnetics.
3. Frankel (1987), in a very useful study of the origins and significance of Hospers' work, points out that his original aim was the geological correlation of lava flows, not Drift or theories of geomagnetism. Hospers' detailed investigations of reversals proved to be of great importance in the development of the programme at Cambridge but he generally interpreted his results as contrary to either polar wander or Drift.
4. Almost every aspect of the programme, from general assumptions to specific techniques, was disputed at one time or another: the origin of the geomagnetic field, the stability of rock magnetism over millions of years, the nature of reversals (*i.e.*, rocks with a magnetism opposite to that of the Earth's present field), criteria for suitable formations, sampling techniques, methods of dating the strata from which the cores had been taken, the possibility of remagnetization by reheating or lightning strikes or other processes which some supposed could alter the original permanent magnetization during the geological history of the strata, magnetostriction, whether thermal or magnetic 'cleaning' revealed or created the 'true' permanent magnetization, and so on.

5. By convention the present orientation of the geomagnetic field in which the 'north' magnetic pole roughly coincides with the north geographic pole is termed 'normal'.
6. The likely source was Gutenberg (1951: 202) which depicted on a map of the north polar region *palaeoclimatic* P-W curves constructed by Koppen and Wegener in support of Drift and polar wander and by Kreichgauer and Milankovitch in support of polar wander. These curves were based on apparent climatic changes as evidenced by geological deposits, a field in which there was considerable controversy and which was often not taken very seriously by more orthodox earth scientists.
7. This serves as one example of how separate images can be juxtaposed and combined into one illustration in such a fashion that the 'whole' is more powerful and persuasive than its 'parts'. Further, the source of Creer's inspiration -- the Gutenberg illustration -- suggests a possible heuristic value in visual images.
8. Creer pointed out (as reported in the 27 September 1954 issue of *Time*) that he was unsure as to whether his results should be explained in terms of Drift or of a general slippage of the entire crust.
9. In appearance, Blackett's group's method of representing results -- in tabular form -- resembled that used by Graham. However, their table represented a very compact summary of extensive fieldwork, the preparation of many samples, the use of special apparatus, thousands of individual measurements, laborious calculations and the application of Fisher's statistics.
10. This was the explanation offered for the agreement the DTM researchers had found between ancient and modern poles: the North American strata that *they* had examined had been overprinted or remagnetized subsequent to their formation and thus were not a reliable indicator of the Earth's magnetism at the time they had been formed.
11. Among those who commented in this vein in textbooks and monographs were: Scheidegger (1958), Dury (1959), Jacobs et al. (1959), Kummel (1961), Goguel (1952t), Read & Watson (1962), L.C. King (1962), Kirkaldy (1962), Longwell & Flint (1962), Zumberge (1963), Scheidegger (1963), Holmes (1965), Ramsay & Burckley (1965), Foster (1966), Read & Watson (1966). As for other specialties, palaeobiogeographers, for example, had reached a near-stalemate over Drift by the 1950s. The data was too unreliable, too fragmentary to conclude on that basis alone whether or not drift had occurred (Frankel, 1981). In the late 1950s and early 1960s specialists in this field, including some opponents of Drift, noted that palaeomagnetic studies had at least re-opened the question (Frankel, 1985, 170-171; 178-180).
12. The reader might repeat an 'experiment' performed by more than a few earth scientists: photocopy or trace the profile onto a sheet of acetate and then fold on the ridge line!
13. The pictures we have examined are a small sample of the rich repertoire constructed and used in the modern revolution. Earth scientists themselves would list among the more famous images excluded here for lack of space Bullard's computer-aided 'fit' of the continents, Tuzo Wilson's origami-like model of transform faults, Heezen and Tharp's portrait of the Atlantic Ocean floor, and Morgan's and Le Pichon's maps of tectonic plates.
14. Several of these original globes survive at the University of Edinburgh, where Creer is Professor of Geophysics.
15. Several manufacturers today produce similar globes for 'demonstration' and teaching purposes. As a variant, the American Geophysical Union markets 'The Puzzle of the Plates' to illustrate plate tectonics, a kit which contains '18 color movable puzzle pieces...*plus* an illustrated, 16 page booklet explaining the theory of continental drift....'

16. Nunan's (1988) examination of the on-going controversy between proponents of plate tectonics and those favouring Earth Expansionism furnishes further examples of the importance of pictorial representations in the formulation of, development of, and controversy over theories. For example, it was S.W. Carey's inability to fit together without gaps or 'gores' stereographic projections of all the continents -- on an Earth of the *present radius* -- which led him to propose that the Earth had had a much smaller palaeoradius and that the Earth had and was continuing to expand. Similar 'map'-based evidence remains a staple argument for Expansionists (Le Grand, 1988: 273).

17. The evolution of palaeomagnetic studies was marked by the invention of new instrumentation and disagreements over techniques. By the early 1960s palaeomagnetics had emerged as a black box: a routinized, unproblematic set of instruments and techniques into one end of which samples were fed and from the other end pole positions were spewed out. This came about through a complex process of negotiation and the resolution of disagreements over a series of theoretical, empirical and methodological issues which ranged from geology into physics, chemistry and mathematics (See Latour, 1987). Many of the instruments and techniques now routinely used by technicians may be the material results of disputes over and eventual consensus on their empirical and theoretical underpinnings. Lakatos (1978: 14-5 & 44-5) acknowledges the theory-ladenness of instruments and Latour and Woolgar (1979: 63-6; also, Latour, 1987) describe instruments as 'reified theory'. See also Giere (1988: 137-40).

18. Block (1981) serves as a non-technical introduction to some major issues, and includes both proponents and skeptics of the roles of and processes involved in mental imagery. Johnson-Laird (1983) argues for the importance of 'mental models' and 'mental images' as part of the cognitive process which is 'hard-wired' in humans. See also in this connection Morris & Hampson (1983). Recent relevant research is presented in Denis *et al.* (1988). Giere (1988: 6, 78-86, 134-7) draws on cognitive science to construct a general account of science.

REFERENCES

Block, N. (1981). *Imagery*. Cambridge, Mass.: MIT Press.

Clegg, J.A., Almond, M. & Stubbs, P.H.S. (1954). The remanent magnetism of some sedimentary rocks in Britain. *Philosophical magazine*, series 7, **45**, 583-98.

Creer, K.M. (Interview). Interview by H.E. Le Grand, 13-15 December, 1988.

Creer, K.M., Irving, E. & Runcorn, S.K. (1954). The direction of the geomagnetic field in remote epochs in Great Britain. *Journal of geomagnetism and geoelectricity*, **6**, 163-8.

Denis, M., Engelkamp, J. & Richardson, J.T.E., eds. (1988). *Cognitive and neuropsychological approaches to mental imagery*. Dordrecht: Martinus Nijhoff.

Donovan, A., Laudan, L. & Laudan, R., eds. (1988). *Scrutinizing science: Empirical studies of scientific change*. Dordrecht: Kluwer Academic Publishers.

Fisher, R.A. (1953). Dispersion on a sphere. *Proceedings of the Royal Society of London*, A, **217**, 295-305.

Frankel, H. (1979). Why drift theory was accepted with the confirmation of Harry Hess's concept of sea floor spreading. In *The reception of unconventional science*, ed. S.H. Mauskopf, pp. 51-90. Boulder: Westview Press.

Frankel, H. (1980). Hess's development of his seafloor spreading hypothesis. In *Scientific discovery: Case studies*, ed. T. Nickles, pp. 345-66. Dordrecht: D. Reidel.

Frankel, H. (1981). The paleobiogeographical debate over the problem of disjunctively distributed life forms. *Studies in history and philosophy of science*, **12**, 211-59.
Frankel, H. (1982). The development, reception and acceptance of the Vine-Matthews-Morley hypothesis. *Historical studies in the physical sciences*, **13**, 1-39.
Frankel, H. (1984). The Permo-Carboniferous ice cap and continental drift. *Comptes rendu de Neuvieme Congres International de Stratigraphee et de Geologie du Carbonifere (1979)*, **1**, 113-20.
Frankel, H. (1985). The biogeographical aspect of the debate over continental drift. *Earth sciences history*, **4**, 160-81.
Frankel, H. (1987). Jan Hospers and the rise of paleomagnetism. *Eos*, **68**, 577-81.
Giere, R.N. (1988). *Explaining science: A cognitive approach*. Chicago: University of Chicago Press.
Glen, W. (1982). *The road to Jaramillo: Critical years of the revolution in earth science*. Stanford: Stanford University Press.
Graham, J.W. (1949). The stability and significance of magnetism in sedimentary rocks. *Journal of geophysical research*, **54**, 131-67.
Gutenberg, B., ed. (1951). *Internal constitution of the earth*. New York: Dover Publications.
Hacking, Ian. (1983). *Representing and intervening*. Cambridge: Cambridge University Press.
Holmes, A. (1944). *Principles of physical geology*. London: Thomas Nelson.
Irving, E. (1988). The paleomagnetic confirmation of continental drift. *Eos*, **69**, 994-1014.
Isacks, B., Oliver, J. & Sykes, L.R. (1968). Seismology and the new global tectonics. *Journal of geophysical research*, **73**, 5855-99.
Johnson-Laird, P.N. (1983). *Mental models*. Cambridge: Cambridge University Press.
Kuhn, T.S. (1970). *The structure of scientific revolutions*. Rev. edn., Chicago: University of Chicago Press.
Kuhn, T.S. (1977). *The essential tension*. Chicago: University of Chicago Press.
Lakatos, I. (1978). *The methodology of scientific research programmes*. Cambridge: Cambridge University Press.
Latour, B. (1985). Les 'vues' de l'esprit. *Culture technique*, no. 14, 4-29.
Latour, B. (1987). *Science in action*. Milton Keynes: Open University Press.
Laudan, L. (1978). *Progress and its problems*. Berkeley: University of California Press.
Laudan, L. (1984) *Science and values*. Berkeley: University of California Press.
Laudan, R. (1982). Tensions in the concept of geology. *Earth sciences history*, **1**, 7-13.
Le Grand, H.E. (1988). *Drifting continents and shifting theories*. Cambridge: Cambridge University Press.
Le Grand, H.E. (1989). Conflicting orientations: John Graham, Merle Tuve and paleomagnetic research at the DTM 1938-1958. *Earth sciences history*, **8**, 55-65.
Lynch, M. (1985). Discipline and the material form of images: An analysis of scientific visibility. *Social studies of science*, **15**, 37-66.
Lynch, M. (1988). The externalized retina: Selection and mathematization in the visual documentation of objects in the life sciences. *Human studies*, **11**, 201-234.
Morris, P.E. & Hampson, P.J. (1983). *Imagery and consciousness*. New York: Academic Press.
Nunan, R. (1988). The theory of an expanding earth and the acceptability of guiding assumptions. In Donovan *et al.* (1988), 289-314.
Phinney, R.A. (1968). *The history of the earth's crust: A symposium*. Princeton: Princeton University Press.

Pitman, W.C. & Heirtzler, J.P. (1966). Magnetic anomalies over the Pacific-Antarctic Ridge. *Science*, **154**, 1164-71.
Raff, A.D. & Mason, R.G. (1961). Magnetic survey off the west coast of North America, 40^0 N. Latitude to 52^0 N. Latitude. *Bulletin of the Geological Society of America*, **72**, 1267-70.
Rudwick, M.J.S. (1976). The emergence of a visual language for geological science 1760-1840. *History of science*, **14**, 149-95.
Rudwick, M.J.S. (1985). *The great Devonian controversy*. Chicago: University of Chicago Press.
Runcorn, S.K. (1956). Palaeomagnetism, polar wandering and continental drift. *Geologie en mijnbouw*, n.s., **18**, 253-6.
Runcorn, S.K., ed. (1962). *Continental drift*. New York: Academic Press.
Shapin, S. (1984). Pump and circumstance: Robert Boyle's literary technology. *Social studies of science*, **13**, 205-228.
Torreson, O.W., Murphy, T. & Graham, J.W. (1949). Magnetic polarization of sedimentary rocks and the earth's magnetic history. *Journal of geophysical research*, **54**, 111-29.
Vine, F.J. & Wilson, J.T. (1965). Magnetic anomalies over a young ocean ridge off Vancouver Island. *Science*, **150**, 485-9.
Wood, R.M. (1985). *The dark side of the earth*. London: George Allen & Unwin.
Wyllie, P.J. (1976). *The way the earth works*. New York: John Wiley & Sons.

University of Melbourne

NOTES ON CONTRIBUTORS

STEPHEN GAUKROGER is Senior Lecturer in the Department of Traditional and Modern Philosophy at the University of Sydney. He studied philosophy as an undergraduate at London University and received his Ph.D. from Cambridge University in 1977. Among his publications are *Explanatory Structures*, *Descartes*, and *Cartesian Logic*. His current interests include seventeenth-century natural philosophy and he is editing a book on seventeenth-century conceptions of antiquity.

JAN GOLINSKI is a Research Fellow at Churchill College, Cambridge, and will soon be taking up an appointment at the University of New Hampshire. He has published articles on the history of chemistry in the early-modern period and on the social history of science. His current research is on public science in eighteenth-century England.

MIRANDA HUGHES is a Ph.D. student in the Department of History and Philosophy of Science at the University of Melbourne. She has published several articles related to her thesis, a study of the Societe des Observateurs de l'Homme and the voyage of Baudin to the southern Pacific.

BRUNO LATOUR is Associate Professor in the Centre de Sociologie de l'Innovation at the Ecole Nationale Superieure des Mines in Paris. Among his publications are *Laboratory Life* (with Steve Woolgar), *The Pasteurization of France*, and *Science in Action*.

HOMER LE GRAND is in the Department of History & Philosophy of Science at the University of Melbourne. A previous life was devoted to technical studies of 18th-century chemistry. His current research in modern geology and geophysics draws upon historical, philosophical and sociological approaches to issues of scientific change and theory choice. Recent publications in this vein include *Drifting Continents and Shifting Theories*.

ANDY PICKERING is in the Department of Sociology of the University of Illinois at Urbana-Champaign, where he teaches courses in the sociology of science and technology. The author of several articles and *Constructing Quarks*, a book on the history and sociology of particle physics, he is currently working through a pragmatist analysis of scientific, and perhaps mathematical, knowledge and practice, of which the present essay is a symptom.

JAN SAPP is in the Department of History & Philosophy of Science at the University of Melbourne. His teaching and research focuses upon the history of modern biology. Recent publications include *Beyond the Gene: Cytoplasmic Inheritance and the Struggle for Authority in Genetics* and *Where the Truth Lies*, a study of Franz Moweus and the origins of biochemical genetics.

JOHN SCHUSTER is Senior Lecturer in the Department of Science and Technology Studies at the University of Wollongong. He has published on Cartesian science, the politics and rhetoric of scientific method discourse, the historiography of the Scientific Revolution, post-Kuhnian models of science dynamics and gender studies of science. His current research includes Cartesian optics, the Latourian model of 'technoscience' and the deployment of the concept of 'paradigms' in recent technology innovation theory.

TERRY STOKES is a lecturer in the Department of Science and Technology Studies at the University of Wollongong. His research focuses on the life scienes. Current projects include patterns of co-authorship in large laboratories and biomedical fraud.

DAVID TURNBULL is a lecturer in Social Studies of Science at Deakin University. His research includes laboratory studies of engineering and parapsychology, gothic cathedrals, maps and meta-analysis. Together with Max Charlesworth and Lyndsay Farrall, he and Terry Stokes are co-authors of *Life Among the Scientists: An Anthropological Study of an Australian Scientific Community.*

GRAEME WATCHIRS lectures in the Department of Science and Technology Studies at the University of Wollongong. His main interest is in the relations between post-structuralist language theories and the sociology of scientific knowledge, with particular reference to the construction of meaning in experiments and the construction of historiographical discourse. He is preparing a textbook on the role of language in science and in theories of meta-science.

INDEX OF NAMES

AUSTRALASIAN STUDIES IN HISTORY AND PHILOSOPHY OF SCIENCE

General Editor:
R. W. Home, *University of Melbourne*

Publications:

1. R. McLaughlin (ed.): *What? Where? When? Why?* Essays on Induction, Space and Time, Explanation. Inspired by the Work of Wesley C. Salmon. 1982 ISBN 90-277-1337-5
2. D. Oldroyd and I. Langham (eds.): *The Wider Domain of Evolutionary Thought.* 1983 ISBN 90-277-1477-0
3. R. W. Home (ed.): *Science under Scrutinity.* The Place of History and Philosophy of Science. 1983 ISBN 90-277-1602-1
4. J. A. Schuster and R. R. Yeo (eds.): *The Politics and Rhetoric of Scientific Method.* Historical Studies. 1986 ISBN 90-277-2152-1
5. J. Forge (ed.): *Measurement, Realism and Objectivity.* Essays on Measurement in the Social and Physical Science. 1987 ISBN 90-277-2542-X
6. R. Nola (ed.): *Relativism and Realism in Science.* 1988 ISBN 90-277-2647-7
7. P. Slezak and W. R. Albury (eds.): *Computers, Brains and Minds.* Essays in Cognitive Science. 1989 ISBN 90-277-2759-7
8. H. E. Le Grand (ed.), *Experimental Inquiries.* Historical, Philosophical and Social Studies of Experimentation in Science. 1990 ISBN 0-7923-0790-9

KLUWER ACADEMIC PUBLISHERS – DORDRECHT / BOSTON / LONDON

www.ingramcontent.com/pod-product-compliance
Lightning Source LLC
Chambersburg PA
CBHW060746150926
51501CB00003B/106
9780792307907